PHYSICS IN THE 20TH CENTURY

PHYSICS IN THE

20TH CENTURY

Curt Suplee

**Edited by Judy R. Franz
and John S. Rigden**

Harry N. Abrams, Inc., Publishers
in association with
the American Physical Society and
the American Institute of Physics

Contents

INTRODUCTION

The twentieth century dawned on an era of exuberant confidence in the power of science and technology to explain the world and to improve the human condition. For scientists and non-scientists alike, that optimism was well founded. The Victorian period had witnessed an incessant parade of scientific and technological astonishments. Thanks to progress in thermodynamics, heat engines—including the newfangled internal combustion engine—had imbued industry with new power. Factories thundered. Railroads roared. Airships cruised the skies. Automobiles hit the road.

Electrical energy, though not yet widespread, promised to transform the nature of daily life. In motors and generators, it would relax the toll on human muscle; in electric lights, it would vanquish darkness itself. Radio waves had been observed for over ten years, telephone communication was becoming possible, and the first Morse code signals were about to cross the Atlantic. Physicists were beginning to understand the nature of that most evanescent substance, light, as a form of electromagnetic radiation. Its speed of 186,300 miles per second had been confirmed to an excellent approximation.

Civilization had every apparent reason to bask in a sense of gathering comprehension of—and thus encroaching dominion over—the objects and energies in the universe. And most educated people shared some of the burgeoning excitement the poet Tennyson felt when he described himself as an "heir of all the ages, in the foremost files of time," content to "[l]et the great world spin forever down the ringing grooves of change."

Beneath it all lay the orderly, predictable, and reassuring cosmos as depicted by Isaac Newton, still imperturbably intact after nearly three hundred years of progress. Newton's classical mechanics—the science of the transfer of forces and the response of matter to those forces—had been refined and extended, but remained the preeminently successful instrument for understanding the way nature worked. By the end of the nineteenth century, writes physics historian Andrew Whitaker, those laws "were held to be among the greatest human achievements, indeed the very greatest strictly scientific achievement, and practically a direct revelation of divine intent." So in 1900, "it seemed unthinkable that they could be challenged."

Nonetheless, strange, unsettling new phenomena were observed in laboratories in Europe and America. Perhaps the most mysterious developments were in the field of radiation, where oddities abounded. Negative electrodes were seen to emit bizarre emanations called "cathode rays." Something called the "X-ray" had just been revealed—a kind of beam that could pass through solid matter almost unimpeded. Equally baffling was the newly discovered process called "radioactivity," in which certain substances appeared to give off an inexplicable form of radiation that could fog a photographic plate.

At the outset of the twentieth century, no scientific topic was more exciting than the newly discovered phenomenon of radioactivity. And no one did more to further the understanding of that mysterious process than Marie and Pierre Curie, shown here shortly before Pierre's untimely death in 1906. Their landmark work—including discovery of the elements polonium and radium—forced a broad rethinking of fundamental principles and made it possible for physicists to investigate the structure of the atom.

Researchers were about to discover that these entities obeyed rules very different from those that governed classical objects such as billiard balls or planets orbiting the sun. In fact, as science began to examine nature at smaller and smaller scales, it soon became clear that a whole new set of concepts would be required to account for a growing body of problems—especially the vexing question of how to determine the energy content of electromagnetic waves. Scientists could begin to explain why a piece of metal placed in a forge glowed red when it emitted long-wavelength radiation; but theory demanded that it should also emit a large amount of radiation at shorter wavelengths, and that

was plainly not the case. And whereas British theorist James Clerk Maxwell had devised an elegant set of equations to describe the wave properties of light, it sometimes behaved in a seemingly unwavelike manner.

Moreover, many scientists still believed that light, like any other wave, needed a medium in which to propagate. Thus the Earth and all other celestial bodies must be surrounded by, and pass through, a hypothetical fluid called the lumeniferous ether. Maxwell insisted that "there can be no doubt that the interplanetary and interstellar spaces are not empty, but are occupied by a material substance or body which is certainly the largest, and probably the most uniform body of which we have any knowledge." However, a cunning and highly accurate experiment conducted by American physicists Albert Michelson and Edward Morley in 1887 found absolutely no evidence of this ghostly medium.

Similarly, although the upstart science of electricity was producing marvels, very little was known about what electricity actually *is*, or what physically happens when a current flows. The electron, nature's chief communicator of negative charge, had been discovered in 1897. But what this "atom of electricity" did, and how it was related to the overall configuration of atoms, was beyond the current state of science.

So was the structure of the atom itself—if, in fact, such things really existed. The hypothetical notion of a single, indivisible atomic unit had been used pragmatically in chemistry to determine the ratios in which various elements combine and statistically in thermodynamics to calculate the motion of large arrays of matter such as gases. But at the end of the nineteenth century, no one was quite sure what atoms *were*, or how they were built. Nor did they understand the physical nature of the bonds that linked atom to atom and gave the periodic table such valuable predictive power. And scientists could not yet even imagine the factors governing the intimate, intricate relationships between atomic behavior and the macroscopic properties of materials—a field that would become the modern discipline of condensed-matter physics.

On the macroscopic scale, there was considerable disagreement over the age of the sun and the Earth. Such grand figures as Hermann von Helmholtz and William Thomson (Lord Kelvin) had theorized that the source of the sun's heat was gravitational collapse of its mass. They calculated that the Earth and its star could not be too much older than 20 to 40 million years.

However, that figure was drastically at odds with the measurements of geologists and paleontologists, who pointed out that some terrestrial surface features such as the Grand Canyon would have taken more than 25 million years to form. But no one could be reasonably certain how long anything had lasted because there was no acceptable benchmark for dating materials of such stupendous ages. And none would be found until physicists began to understand how radioactive elements break down over precise and predictable time scales.

But of all the subjects of physical science, perhaps none epitomized the late-nineteenth-century paradox of enormous progress and lingering incomprehension as much as the cosmos. An intelligent, well-educated person gazing up at the night sky in 1900 would have known about planetary orbits and other gravitationally bound systems around our sun, a bit about comets, and a fair amount about ways astronomers characterized and catalogued visible objects in the heavens.

No one had begun to understand why stars shine, or how they are arranged in the heavens, or how far away they are. In fact, there was no suitable approximation of the size of the universe (though some adventurous minds had postulated its scope at a few light years), or any idea of how it might be moving or evolving on grand scales. Many observers believed that our galaxy, the Milky Way, *was* the universe.

Within a few decades, all of that would change forever, as revolutionary discoveries altered thousands of years of human history in an accelerating expansion of knowledge that continues to this hour. At the largest scales, Einstein's theory of relativity utterly destroyed the traditional views of space and time. Space, the world learned to its shock, warped and buckled in the presence of mass. Time was not a universal constant; from Earth's perspective, it ran at completely different rates in different locations.

At the smallest scale, traditional thinking turned out to be practically useless. Exploring the uncharted realm of matter and energy at atomic dimensions, physicists found that nature follows an entirely unexpected and wholly different set of principles from the familiar Newtonian systems. Particles—unlike things in the dependable, predictable world of everyday objects—do not have a definite position or other characteristics. In fact, many physicists would argue, they do not "exist" in the conventional sense at all unless they are observed!

Matter can become energy; energy can turn into matter. Light is neither a particle nor a wave; it is both, or either, at different times. Even the emptiest vacuum is seething with activity as particles pop into and out of existence. Elements transform themselves into other elements by splitting or fusing, releasing energy of unimaginable magnitude. Electrons travel through materials according to complex but comprehensible causes and sometimes do so without encountering any resistance.

As the new knowledge of atomic and subatomic entities grew, scientists found ways to apply it to objects as small as the circuits on computer chips and as large as the violent death spasms of giant stars. As early as 1905, Ernest Rutherford would marvel, "The rapidity of this advance has seldom, if ever, been equaled in the history of science."

In 1999, those sentiments are just as accurate. The world as it is now understood differs more dramatically from the Victorian view than the science of Galileo, Copernicus, and Newton differed from that of Aristotle. By the end of this century of wonder, physics would transform nearly every aspect of daily life. It would also permanently alter our vision of reality, the universe we inhabit, and our peculiar and enigmatic place in it.

1 ATOM [10]

Today, we feel quite at home in an atomic world. We take it for granted that all the stuff we encounter in everyday life is made of trillions upon trillions of individual units called atoms, and that they in turn contain even smaller components in the form of electrons, protons, and neutrons. Those ideas now seem so comfortable and fundamental that it's hard to imagine that they were considered suspiciously radical in our grandparents' lifetimes. Indeed, the discovery and characterization of atoms is perhaps the preeminent triumph of twentieth-century physics.

"Atomism" was not a new concept. Greek sages had posited it in the fifth century B.C., and by the early nineteenth century, British scientist John Dalton had become convinced that matter was built from tiny entities that were "absolutely indecomposable." In addition, chemists had learned that elements combine only in certain specific ratios by weight—a notion implying that each element must occur in discrete units.

Yet at the dawn of the modern era—with automobiles, telephones, and radios already in widespread use—there were distinguished scientists who doubted that atoms had a real, physical existence. After all, nobody had ever seen one. (And for good reason: as physicists later determined, even a fairly hefty atom is about 10^{-8} meters wide, about $1/10,000$ the width of a human hair. And 99.9 percent of its mass is in the nucleus, which is 10,000 times smaller yet!) So at the turn of the century even such formidable figures as Austrian physicist Ernst Mach were still insisting that the supposed atom was no more than a useful fiction.

Within a few wondrous decades, however, scientists had not only revealed the structure and behavior of the atom in exquisite and astonishing detail, but were using that knowledge to understand natural phenomena on scales from the submicroscopic to the cosmic.

As if symbolizing the energy and excitement physicists felt at the beginning of the twentieth century, electrical discharges can be seen arcing across Nikola Tesla's laboratory in Colorado Springs, Colorado. This multi-exposure photograph, taken in 1889, shows Croatian-born Tesla sitting beside the electrostatic generator he designed and built. In the ensuing decades, scientists devised increasingly sophisticated ways to probe matter at unprecedented energy levels.

Electrons

The first convincing clue was the discovery of the electron—or, as it was then known, the "cathode ray." In the mid-nineteenth century, scientists had found that if electrodes were placed in a vacuum tube, the negative pole, or cathode, appeared to emit some strange form of radiation. By the 1890s, voltage generators had become powerful enough, and vacuum conditions good enough, that this effect could be observed in detail. But no one knew what it *was.*

Many Continental researchers were betting that it was indeed radiation. German physicist Heinrich Hertz had shown that cathode rays could penetrate a thin metal foil, which was very ray-like behavior. Moreover, X-rays and radioactivity had just been discovered, and mysterious sorts of radiation suddenly seemed to be cropping up almost monthly. But some British physicists had suspected for years that the rays were actually streams of an unknown kind of particles carrying negative electrical charges.

One of the physicists was Joseph John (J. J.) Thomson, son of a Manchester bookseller, who in 1884 had been elected as director of Cambridge University's famed Cavendish Laboratory. In the mid-1890s, he set out to examine the phenomenon using a then-high-tech apparatus designed by Sir William Crookes. The device was in many ways no different from a modern television set: in a sealed glass tube from which most of the air had been pumped to create a vacuum, the cathode emitted its rays in straight lines that made sections of the glass near the cathode glow brightly, just as a TV "electron gun" shoots particles at a glass screen coated with phosphors.

Thanks to the laws governing the interaction of charges and fields—worked out in the nineteenth century by British physicists Michael Faraday, James Clerk Maxwell, and others—Thomson knew that if cathode rays were actually streams of charged particles, they should be deflected by electric and magnetic fields. The direction in which the beam was bent would reveal the type of charge, and the amount of the deflection would depend on the size of the charge and the speed of the particles.

Previous attempts to demonstrate this effect had failed. But Thomson surmised that the fields might have been too weak. Using improved induction coils and a better vacuum, he found that he could bend the beams, causing the glow to shift to a different part of the tube. In 1897, he wrote, "I can see no escape from the conclusion that they are charges of negative electricity carried by particles of matter." He had also been able to determine the ratio of the mass to the charge. Though he could calculate neither quantity separately, it appeared likely that the mass was shockingly scant—around 1/1,000 that of the positively charged hydrogen ion, which had been approximated by chemists and was "the smallest mass hitherto recognized as being capable of a separate existence."

(More than a decade would pass before American physicist Robert Millikan was able to determine the new particle's charge accurately. First he noted how fast tiny oil droplets fell in air. Then he induced an electric charge on the droplets that would push them in the opposite direction and measured how much charge was necessary to propel them upward. His data came out in multiples of a single, presumably minimal charge. In

1909, he determined this quantity to within a few percentage points of the currently accepted value. Once he had established the charge, he could calculate the mass, which would prove to be 1/1,837 that of the hydrogen nucleus.)

Thomson had called his particle a "corpuscle." But the name that stuck had been invented in 1891 by his Irish contemporary G. Johnstone Stoney: "electron." Its discovery jolted science into a new way of imagining the composition of matter—and suddenly raised several serious questions.

For one thing, if atoms existed, they were supposed to be the smallest elementary units of matter. But here was something thousands of times smaller. And it couldn't be the atom itself. Scientists had long known that, if there were such things as atoms, they were electrically neutral, although they could be made to take on a positive charge (that is, become ions) if exposed to enough energy. Today, we understand that the energy dislodges one or more electrons, leaving the ionized atom with a net positive charge from the protons in its nucleus.

At the end of the nineteenth century, however, all Thomson knew was that if the electron was negative, something in the atom had to carry a corresponding positive charge. But what was it? And how were the charges arranged?

Various ingenious models were proposed. But early in the twentieth century the favored conception was the one endorsed by Thomson: an atom was a composite in which a number of electrons are imbedded in a wad of undifferentiated positive matter "like raisins in a pudding." This agreeable notion didn't last long.

This is the device built and used by Robert Millikan in 1913 to determine the electrical charge of the electron. By measuring the speed with which electrically charged droplets of oil fell through electric fields of various strengths, Millikan arrived at a result that differs by less than 3 percent from the current accepted value.

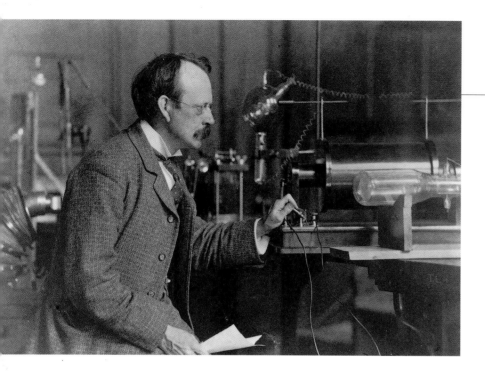

British physicist J. J. Thomson ushered in a new era of subatomic physics by discovering the electron in 1897. Here he is shown in the 1890s with the apparatus he used to determine the ratio of the electron's electrical charge to its mass.

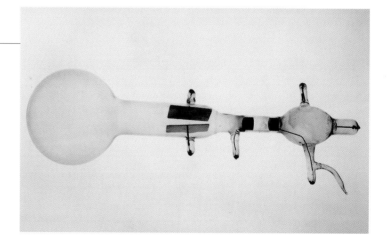

This device, designed by Thomson, was crucial to his experiments. Air was pumped out of the glass enclosure so that electrons (then known as "cathode rays") would not collide with gas molecules. The negatively charged cathode at the far right generated electrons that were accelerated toward a positively charged plate with a slit in it. Some electrons passed through the slit, and proceeded at about one-tenth the speed of light toward the large bulb at the left, where the beam made a luminous spot when it collided with the glass at far left. As the electrons passed by two more electrodes, mounted on the top and bottom of the glass just before the large bulb, they were deflected from side to side. The amount of deflection depended on the voltage between the two electrodes. By applying a magnetic field from outside coils (not shown) that exactly balanced the deflection, Thomson was able to find the essential ratio between the charge and mass of the electron. Once the charge was known, the mass could be calculated.

A century after Thomson's experiments, the successor to his cathode-ray bulb is the modern particle detector, which takes millions of measurements in a fraction of a second to determine the mass, charge, and other characteristics of various particles. This detector, shown in construction at the Stanford Linear Accelerator Center in California, surrounds the open particle beam collision chamber at the center.

Nucleus

While Thomson was examining his corpuscles, researchers on the Continent were pondering another new phenomenon—radioactivity. First observed in uranium by French physicist Henri Becquerel in 1896, it was initially inexplicable. Certain substances gave off emissions that left an image on a photographic plate. But what was being emitted?

The first comprehensive answer was provided by Ernest Rutherford, a bluff and hearty New Zealander who worked under Thomson at Cambridge before moving to McGill University in Montreal, Canada, and then back to England at the University of Manchester. By the turn of the century, Rutherford had examined uranium emissions and determined that there were two very different kinds.

One, which he called alpha, was very easily absorbed by materials as thin as a piece of paper. The other, beta, was much more penetrating and could much more easily pass through a thin sheet of aluminum. (Becquerel would later confirm that these beta rays were actually electrons.) Shortly after he arrived at Manchester in 1907, Rutherford had concluded that alpha particles were positively charged because of the way they were deflected by electric and magnetic fields. The relatively small amount of that deflection suggested a sizable mass. He eventually decided that the alpha particles "must consist of atoms of helium"—an element whose mass was known approximately from chemistry—and ejected from certain unstable elements at about one-twentieth the speed of light.

We know now that the alpha particle consists of two protons and two neutrons (precisely the same as the nucleus of the most common isotope of helium), which explains its positive charge. At the time, however, all Rutherford could discern was that these comparatively massive, high-energy particles could be stopped or greatly diverted by nothing more than a tissue-thin sheet of metal—suggesting that there were some kind of diminutive but impenetrable barriers lurking in matter. Using radioactive material as a sort of mini-cannon for firing alpha particles at a piece of sheer gold foil, Rutherford and his associates set out to study what happened to the particles as they collided with the gold film.

By that time other researchers had devised a cunning method of tracking such trajectories: a charged particle striking a layer of zinc sulfide would produce a burst of light, or "scintillation," at the impact point. Rutherford's team arranged such a layer near the target. By observing the location of the light bursts carefully with a microscope, they began to get a good count and a very accurate measure of the small angles through which the alpha particles were deflected.

And so they might have continued. But one day in 1909, Rutherford's research assistant Hans Geiger (inventor of the eponymous radiation counter) teamed up with Rutherford's student, Ernest Marsden. Rutherford suggested that they see whether any particles were scattered through relatively large angles. It didn't seem a terribly promising idea, but it would transform physics.

New Zealand–born physicist Ernest Rutherford discovered the nature of the atomic nucleus, among numerous
other landmark accomplishments. He is shown here in his
laboratory at McGill University in Montreal in 1905.

To everyone's amazement, Geiger and Marsden found that one in every few thousand of the particles bounced back out of the foil in the general direction of the source, as if they had collided with something solid inside. "It was as incredible as if you had fired a fifteen-inch shell at a piece of tissue paper and it came back and hit you," Rutherford was later fond of saying.

The Thomson pudding model of the atom simply could not account for this kind of behavior because it was supposed to be homogeneous. Rutherford became convinced that the atom was not uniform in density. It must contain a tiny "nucleus" of matter about 1/100,000 the size of the whole atom. That solid core, on the order of a few hundredths of a trillionth of a meter wide, must contain something with a positive charge strong enough to repel the occasional alpha particle that came close to it. Nearly all of the atom was nothing but empty space!

So what *was* in the nucleus, and what was carrying the positive charge? To explore that question, Rutherford and Marsden switched to smaller target atoms. Gold was so massive, and its nucleus presumably had such a large positive charge, that the modest alpha particle was unlikely to knock anything out of it. But atoms in various lightweight gases such as hydrogen and nitrogen might be much more susceptible. And so they proved to be.

In 1914, when Marsden bombarded air with alphas, he dislodged a strange, positively charged object that was several times less massive than the helium nucleus. Pursuing this observation intermittently during World War, I by early 1919 Rutherford had found proof that "H particles," which he soon named "protons," could indeed be dislodged from the nitrogen nucleus. The "H" particle had the same characteristics as the hydrogen ion, which had been extensively studied. This suggested that atoms of nitrogen contained the same kind of basic particles found at the core of hydrogen. The nuclear proton appeared to be the electrical counterpart to the electron, with an identical but opposite charge.

The atom was becoming comprehensible. But like all truly profound revelations in science, the new findings raised troubling new questions. For example, how, exactly, were electrons bound to atoms?

Rutherford and others posed a comprehensible, if extraordinary solution: electrons must orbit the nucleus like planets around the sun. But that model had a fatal weakness. Electrons are charged particles. And Maxwell's laws demanded that if such a charge were traveling in a circular orbit, it would give off energy as electromagnetic radiation in much the same way that a radio antenna emits waves. If that were the case, the electrons would very quickly dissipate their energies, slow down, and plunge into the nucleus. Some process beyond classical physics, it seemed, prevented this from happening. But what?

The man who confronted that quandary was Niels Bohr, son of a Copenhagen physiology professor. Bohr had worked briefly in Cambridge with Thomson and then went to Rutherford's lab at Manchester, where in 1912 he began to devise a revolutionary picture.

Bohr retained the idea of orbits, but rejected the classic idea that the electron would radiate continuously. Instead, he turned to a bizarre and then somewhat disreputable concept posited in 1900 by German physicist Max Planck.

Planck had argued against the entirely sensible assumption that the energy emitted by excited atoms comes in a smooth, unbroken gradation of values. Experimental observations of heated bodies could be better explained, he determined, if matter gave off energy in discrete units that would be called "quanta" from the Latin word for "how much." The amount of energy in each discontinuous quantum was a function of its frequency.

Similarly, Bohr reasoned that electrons could encircle atoms only in certain allowed orbits at particular distances from the nucleus. Each permitted orbit corresponded to a slightly different energy, and electrons only emitted or absorbed radiation when they changed from one of those energy conditions, or "levels," to another. Further, once an electron was at the lowest possible energy, called the "ground state," it could not radiate at all.

(As we will see in the next chapter, today we go one step further and accept that when an electron drops from a higher to a lower energy level, radiation in the form of light quanta is emitted. In 1905, Einstein described the photoelectric effect by postulating that light is made up of quantized energetic particles. At the very beginning of the century, however, that notion was so completely outlandish that physicists, including Bohr, did not generally accept it. Nearly two decades would pass before experiments by American physicist Arthur Holly Compton confirmed the notion and made Einstein's "corpuscular concept" widely credible.)

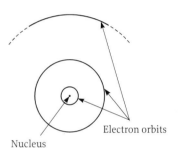

Nucleus

Electron orbits

The Bohr atom was not convincing when first presented in 1913, but it made excellent sense of a phenomenon that had baffled scientists for centuries: spectra. When an element is heated, it gives off radiation that can be split into its component wavelengths using prisms or gratings to spread out the pattern. By the nineteenth century, it was evident that each element had its own unique spectrum with a telltale pattern of lines: black ones marking wavelengths absorbed by the substance, and bright ones representing the wavelengths emitted.

But there was no suitable explanation for *why* this happened, although in 1885 Swiss mathematician Johann Balmer had devised a formula that neatly described the relationship among the wavelengths that made up the major lines in the visible hydrogen spectrum (called the Balmer series). As it turned out, Bohr's theory exactly predicted their placement. To many, it seemed that the atom at last was comprehensible.

The jubilation was premature. Bohr's model failed to work on more complicated atoms—a situation that spurred scores of researchers to seek, and eventually find, utterly unexpected new characteristics. Nonetheless, by 1913 the general outline of the modern atom was becoming clearer. The conviction was spreading that a new age of physics had arrived. It would unfold as investigators learned to use ever more sophisticated means to probe deeper and deeper into the secrets of matter.

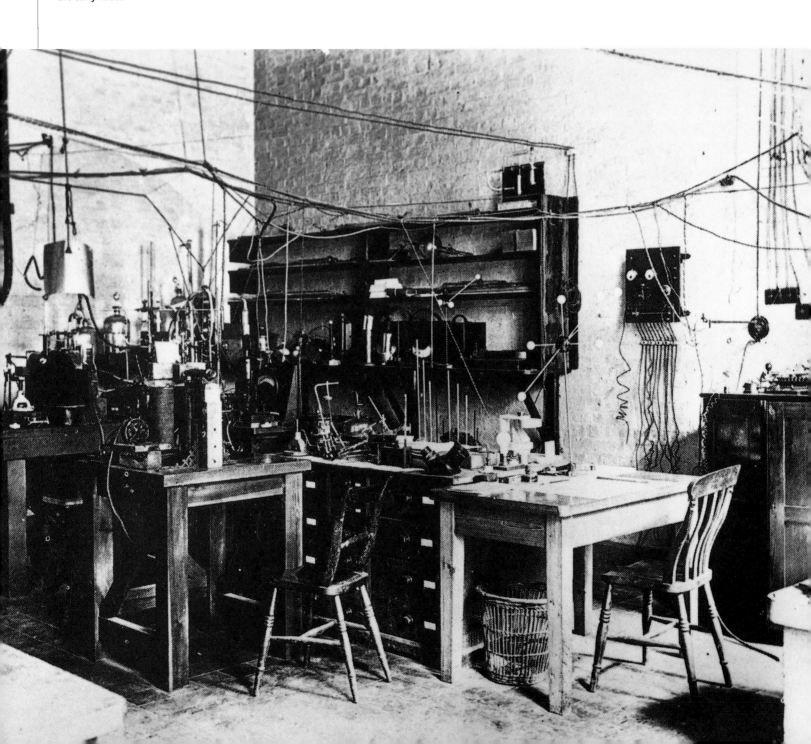

Rutherford succeeded J. J. Thomson as director of the famed Cavendish Laboratory at Cambridge University. Shown here is his research facility in the early 1920s.

Modern researchers investigating the composition of matter and other physical systems use many techniques. One involves bombarding samples with beams of high-energy X-rays or ultraviolet radiation. These beams are generated by particles accelerated in a circular pathway called a synchrotron. In this apparatus at the National Synchrotron Light Source at Brookhaven National Laboratory on Long Island, beams are extracted at various points around the particle loop, and physicists arrange their instruments in line with the beam.

On the Firing Line

It would not be easy. Much of physics involves poking objects with something and observing the result. But in the subatomic realm, there are relatively few objects small enough to do the poking. Rutherford's alpha particles were far too large—and insufficiently energetic—to reveal the fine structure of the atom. "If alpha particles—or similar projectiles—of still greater energy were available for experiment," Rutherford wrote in 1919, "we might expect to break down the nuclear structure of many of the lighter atoms."

Projecting lightweight electrons from a cathode was one thing. Boosting the energy of a proton high enough to disintegrate a nucleus was quite another. It would require what in the 1920s seemed unattainable electrical potentials: millions of electron volts (eV). (One eV is the energy an electron acquires when accelerated across a potential difference of 1 volt.)

Nonetheless, two researchers at the Cavendish Laboratory, John Cockcroft and Ernest Walton, set out to see how close they could get. Using a high-voltage transformer, they began accelerating protons down a tube between the two oppositely charged plates at energies up to 750,000 eV. To their delighted surprise, in 1932 they succeeded in banging a proton into an atom of lithium (the third-lightest element) so hard that its nucleus absorbed the proton and split into two separate nuclei of helium, the second-lightest element.

Meanwhile, on the other side of the Atlantic, a young American physicist named Robert Van de Graaff had invented an even more powerful device. It used an insulated conveyor belt to carry positive charges to the interior of a hollow metal dome. The longer the belt ran, the more voltage built up in the dome. Positively charged hydrogen ions—that is, protons—placed inside the dome were repelled by its powerful electric field and propelled out of the apparatus through an accelerator tube to energies of 1.5 million eV.

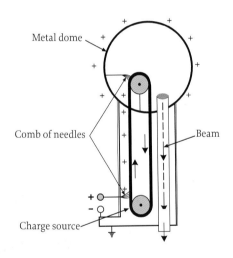

Metal dome

Comb of needles

Beam

Charge source

The Cockcroft-Walton and Van de Graaff machines conferred energy to particles in a single stupendous blast. Thus they were inherently limited by the amount of charge that could be stored. But a Norwegian engineer, Rolf Wideröe, had designed a system whereby drastically greater amounts of energy could be conveyed to particles by applying relatively small accelerations, but doing it many times.

In Wideröe's scheme, a charged particle would proceed down the center of a series of metal tubes separated by short gaps. Just as the particle reached the gap, the tube in front of it would be given the opposite electrical charge, which would attract it. Thus at each gap, the particle would accelerate. The limiting factor was size: Kicking particles up to very high energies would require an extremely, and perhaps prohibitively, long line of tubes.

In 1929, a young Berkeley physics professor named Ernest Lawrence was studying Wideröe's ideas and realized that the process could be made circular, forcing the particles to pass across the same accelerating gap again and again, by exploiting a well-known principle. That is, a charged particle moving through a magnetic field experiences a sideways force that causes it to begin to rotate in circular fashion.

Particle accelerators are required to probe the nucleus and learn its properties. One type of particle accelerator was invented by Robert Van de Graaff in 1929. The Van de Graaff accelerator, which uses principles of electrostatics to accelerate charged particles, is shown here, with its inventor, in its 1931 version. The Van de Graaff accelerator is still used in nuclear physics.

Ernest Lawrence invented another type of accelerator, called the cyclotron, in 1931. Here Lawrence holds his first cyclotron in the palm of his hand.

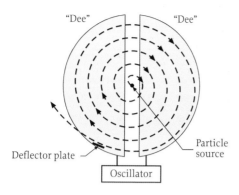

"Dee"   "Dee"

Deflector plate

Particle source

Oscillator

Lawrence's "cyclotron" design had two semicircular, D-shaped hollow metal chambers, called Dees, arranged with their straight sides facing each other—but separated by a small gap. He then put the Dees in a magnetic field. A charged particle in the device would start to revolve. When it reached the gap, the charge on the Dee it was approaching would be made opposite to the particle's, accelerating it across the gap. After it had gone halfway around, the voltage would be reversed: the particle would be repelled by the chamber it was leaving and attracted by the chamber it was approaching. The key was to make the timing of the voltage change match the time it took the particle to complete half of its circuit.

Smashing Success

Lawrence's prototype, built in 1931, was just 5 inches in diameter, yet it accelerated protons to a whopping 80,000 eV. Larger models quickly followed that reached millions of eV. The early machines benefited from a happy fact of physics. Each time a particle was accelerated in the cyclotron, the radius of its trajectory increased slightly and it traveled through a larger arc; but since it was moving faster, it completed one circular lap in exactly the same amount of time. Thus the timing frequency of the voltage change could remain constant as the stream of particles gathered more and more energy. But as the energies got higher, Lawrence soon ran into one of nature's few insuperable barriers: the speed of light.

As a particle approached light speed, it began to behave in strange ways predicted by Einstein decades earlier. He had postulated that any entity with mass could never attain that speed (designated c) because as it got close, its mass would start to increase. Lawrence's team began to see that effect in the mid-1940s. When the particles reached orbits over 10 feet in diameter and speeds of 0.2 c, they began to grow so heavy that they took slightly longer to complete each circuit, spoiling the timing.

The solution, physicists found, was to synchronize the frequency of the accelerating field with the circulating particles to compensate for relativistic deceleration. There was a price to pay: accelerators could no longer contain a continuous stream of particles at different energies, as the original cyclotron had; they could propel only one bunch of particles, all at the same energy.

But the new "synchrotrons" would eventually boost their projectiles to within a whisker of the speed of light, attaining energies of billions and even trillions of electron volts in rings that were miles in diameter. Collisions in those behemoths would reveal the inner structure of the atom in dazzling and unprecedented detail.

Thus, within the course of a few brief decades, science went from corpuscles, "H-particles," and raisin puddings to a highly sophisticated understanding of the atom and its components. That, in turn, enabled physicists to devise a striking array of investigative techniques that would speed the process of discovery and lead to dozens of practical inventions that revolutionized biology, chemistry, and medicine. Among them were the electron microscope, the scanning tunneling microscope, and nuclear magnetic resonance.

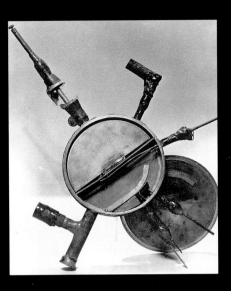

The original 1931 cyclotron, shown here, was about four inches in diameter. At the center was a source of electrically charged particles, or ions. Above and below the device were magnets (not shown) whose field extended through the cyclotron. Charged particles moving in a magnetic field experience a force that pushes them at right angles to their direction; as a result, they begin to rotate in a circular path, passing alternately through each of two D-shaped metal chambers, called Dees. An alternating voltage is applied to the Dees. As particles pass from one side of the cyclotron to the other, the charge of the Dee they are exiting is given the same charge as the particle, which pushes it away. The Dee that the particles are entering is given the opposite charge, which attracts them. Thus, on each circuit the particles gain more energy, and their radius of motion gets larger. Finally, after the spiral journey has raised them to very high energies, the particles escape through the outermost section of the Dees.

Modern particle accelerators are gargantuan compared to their forerunners. In this aerial view of the Fermi National Accelerator Laboratory in Illinois, the circle, approximately four miles in circumference, identifies the path followed by particles as they are accelerated to speeds approaching the speed of light.

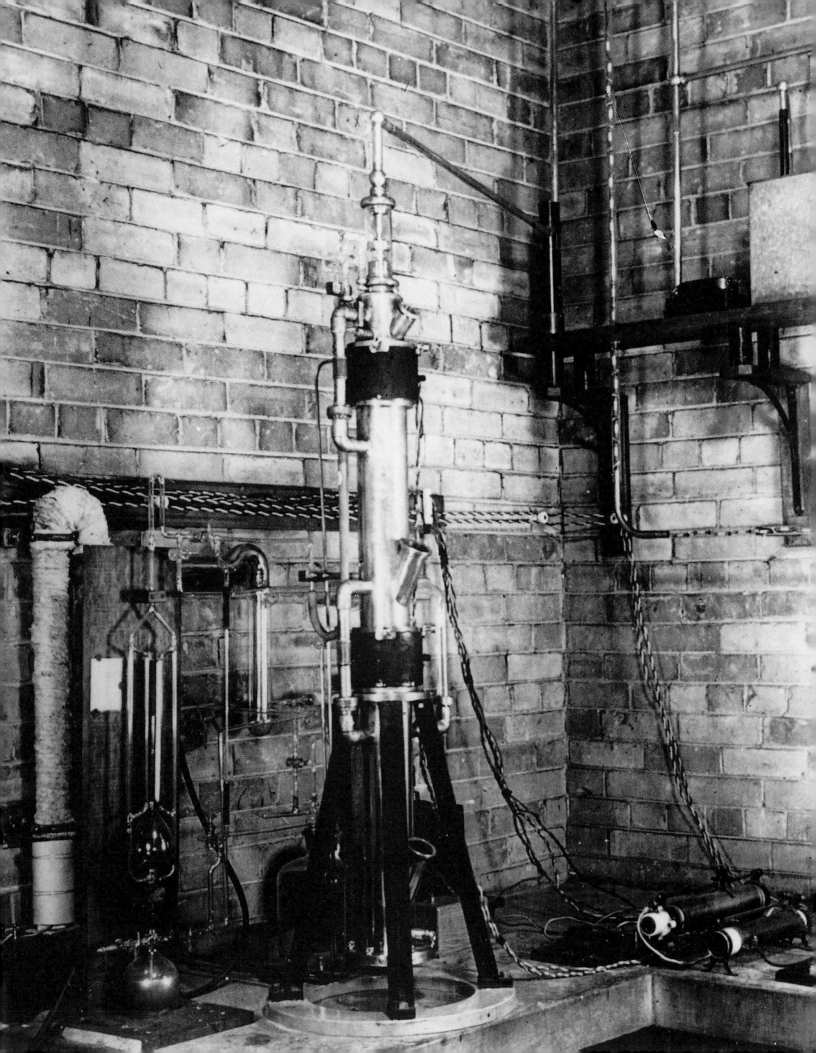

The first electron microscope looks rather crude when compared to a commercial offspring. This instrument combines a scanning tunneling microscope, a scanning electron microscope, and a scanning Auger microscope and is used for high resolution structural and chemical analysis.

Microscopes without Light

Investigating matter on the tiniest scales has an inherent problem: it is impossible to get an image of something smaller than the smallest units used to view it. Thus waves of visible light (which average around 500 nanometers—nm, or billionths of a meter) are too big to reveal even a sizable atom, which is about 0.2 nm wide, or indeed anything much smaller than a few millionths of a meter.

But in the early 1920s, as we shall see later, French theorist Louis de Broglie had made the then outrageous argument that matter had wave-like properties, and that the wavelength of a particle got shorter as its velocity increased. Soon researchers discerned that electrons could be used to observe very small structures, since an electron accelerated by tens of thousands of volts has a wavelength on the order of 0.005 nm—around 100,000 times smaller than visible light.

The first electron microscope built and successfully operated in North America was built in 1938 in the Department of Physics at the University of Toronto under the direction of E. F. Burton.

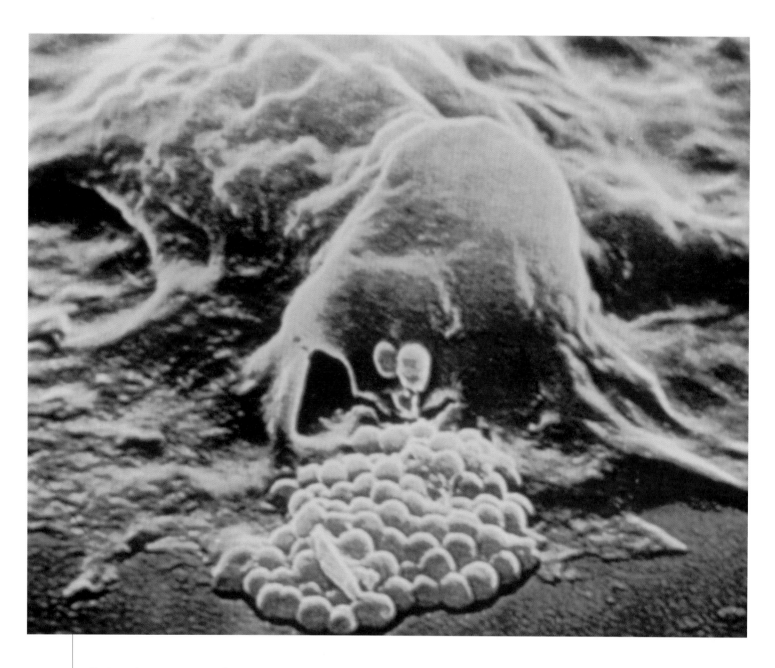

Electron microscopes can reveal great detail in both physical and biological objects. Here a scanning electron microscope captures a white blood cell, called a phagocyte (from the Greek for "eating cell"), in the process of pursuing an invading bacterial colony. Phagocytes, which constitute part of the human immune system, are capable of engulfing other microbes or bits of damaged tissue.

The electrons would have to be focused with magnets. Just as a glass lens can bend visible light, a magnetic field could act as a lens for high-energy electrons. By 1932, German physicist Ernst Ruska had built the first "electron microscope" that was more powerful than its optical counterpart. The next year, the first commercial models went on sale. By the end of the century, resolution was so fine that electron microscopes could be used to create images of individual large atoms.

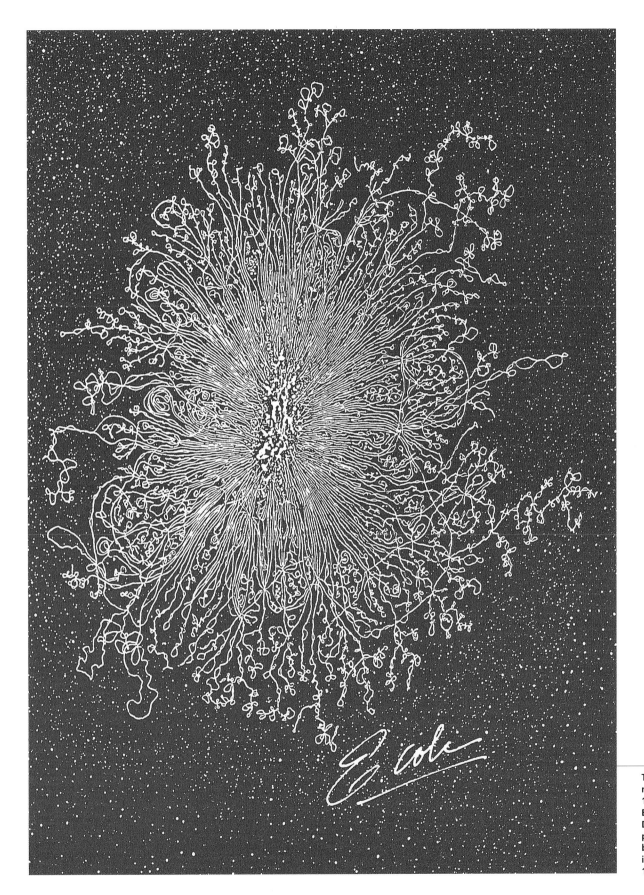

This transmission electron microscope image made in 1983 by Ruth Kavenoff and Brian Bowen shows one DNA molecule in a genome purified from the intestinal bacterium *E coli*. The DNA is magnified by a factor of 15,000.

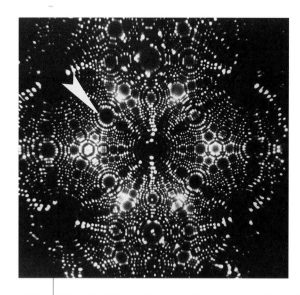

Physicists are now able to obtain images of individual atoms in a number of different ways. These images were made by computer reconstruction of signals from a device with a magnifying power over 2 million called an atom-probe field ion microscope, developed by Erwin Mueller at Pennsylvania State University in 1955. Each of the dots in the image on the right represents the location of a single tungsten atom on a tungsten surface. In the image on the left, the operator has selected a single atom (marked by the arrow) for analysis.

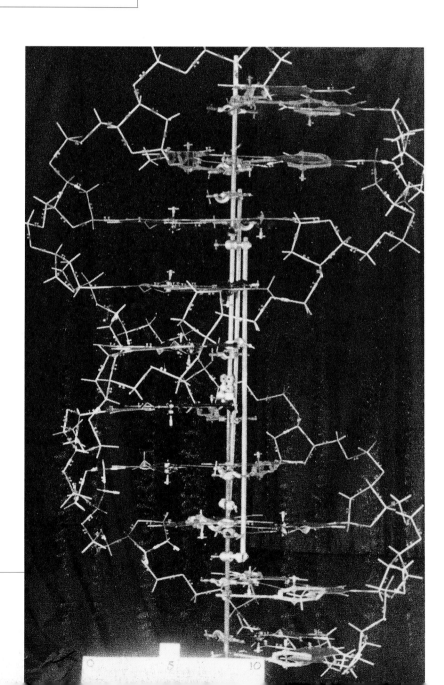

The combination of numerous techniques from modern physics has made it possible for investigators to understand the chemical structure of even highly complicated organic molecules. X-ray crystallography, further discussed in Chapter 2, is an important tool. In 1953, James D. Watson and Francis Crick determined the structure of DNA after analyzing X-ray diffraction photographs made by Rosalind Franklin. This is their model, built at the time to show the double helix structure of the molecule.

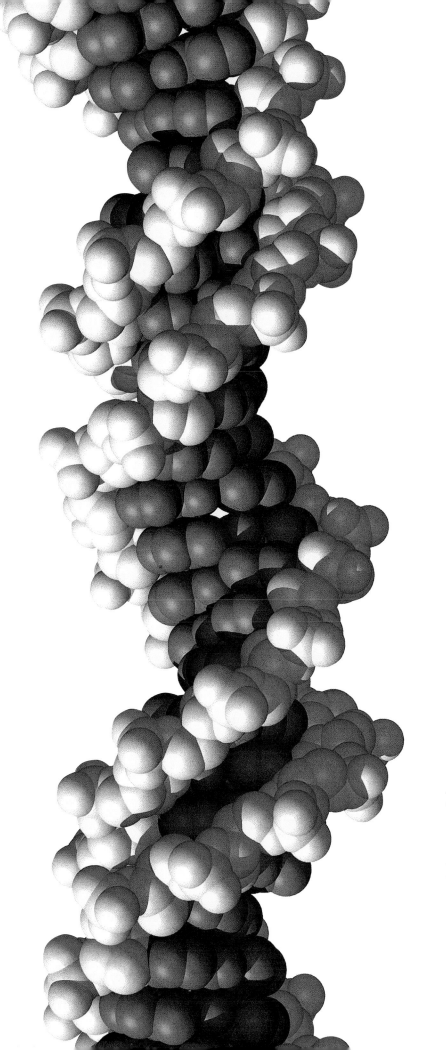

This is a contemporary computer representation of the structure of DNA. The two phosphate-sugar strands that make up the famous "double helix" (white) are bound to each other by links (black) tying together groups called bases.

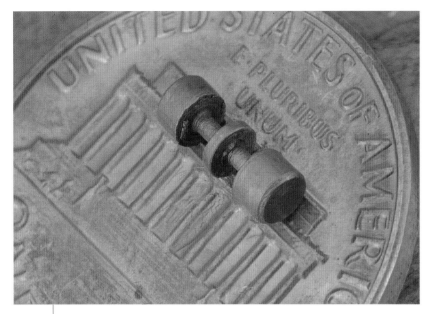

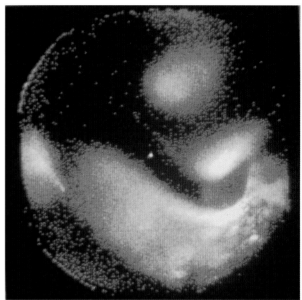

Neutral atoms become ions by either the addition or loss of electrons. Due to their electrical charge, ions can be "trapped" and immobilized by electric fields. The electrode trap pictured here (shown atop a penny to indicate scale) captures mercury ions. The high-magnification photo to the right—taken with a video camera that can record ultraviolet light—shows a single trapped mercury ion (the glowing dot at the center). Such individual ions are held in place by electric and magnetic fields; otherwise, the trapped ions are isolated from outside influences. Thus they can be used to make highly accurate atomic clocks.

Handling Atoms

It took somewhat longer for another quantum-mechanical concept—called tunneling—to work its way into a practical imaging device. Shortly after de Broglie advanced his notion of matter waves, Austrian physicist Erwin Schrödinger invented a landmark equation that described one of the weirder aspects of matter at the subatomic level: matter does not actually exist in one definite place or condition; instead, it has a certain probability of existing in a variety of places and conditions.

In the case of a single electron, that means that even if it is confined inside solid matter, there is a small but real possibility that it can leak outside and enter another solid that is very close by. This seemingly magical (but strictly natural and mathematically explicable) ability is called "tunneling."

One of its many uses was conceived in 1981 when two scientists at IBM's Zurich research center, Gerd Binnig and Heinrich Rohrer, set out to investigate the effect. They made a needle whose tip was only a few atoms thick. When they moved it over a sheet of gold—no more than a couple of atoms' width from the surface—they could detect the tunneling current as it moved from individual atoms to the needle tip. Their "scanning tunneling microscope" (STM) could easily distinguish one atom from another and map the terrain of solid surfaces in fantastic detail.

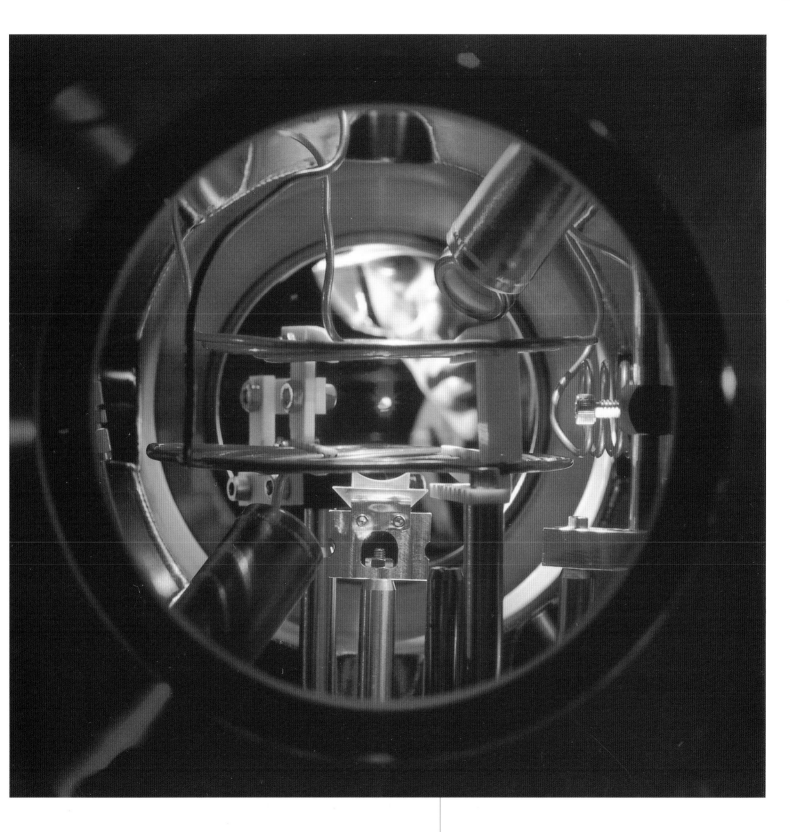

This device, a magneto-optical trap from the U.S. National Institute of Standards and Technology, holds a cloud of sodium atoms, which can be seen glowing in the center of the apparatus. While they are held stationary by magnetic fields, the atoms are gently buffeted with photons; and as the atoms lose their energy to those photons, they cool. This cloud has a temperature of a few tenths of a millionth of a degree above absolute zero (-273.15 C). The physics of atom traps is discussed in Chapter 3.

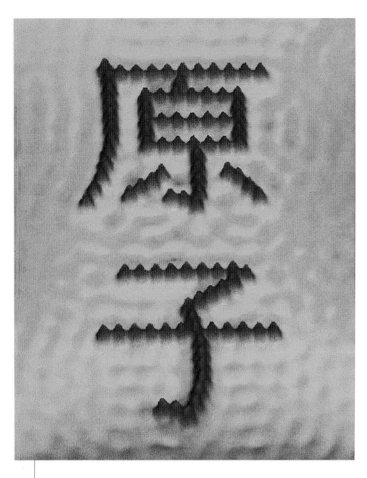

With the scanning tunneling microscope, atoms can be individually imaged and manipulated. In this image, produced by members of the IBM Research Division, iron atoms are positioned on a copper base to form the Japanese Kanji characters for "atom."

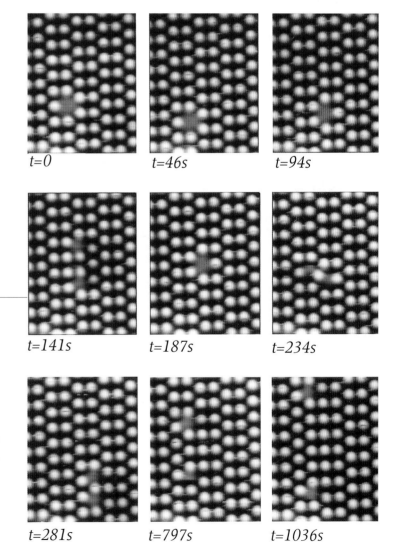

t=0 *t=46s* *t=94s*

t=141s *t=187s* *t=234s*

t=281s *t=797s* *t=1036s*

At the end of the twentieth century, physicists are able to study the behavior of atoms in exquisite detail. These images of a germanium crystal surface, prepared by scientists at France's Commission for Atomic Energy, show how the position of a vacancy (missing atom) in the lattice changes over time. After a single atom is removed from the crystal, it leaves a hole. At room temperature, germanium atoms have enough thermal energy that other atoms can "hop" into the vacated space, leaving a hole elsewhere. Over time, the hole seems to wander around the crystal surface. In this sequence, "t" stands for time and "s" for seconds.

A scanning tunneling microscope uses infinitesimal differences in current moving between individual atoms and a tiny needle tip to reveal the topography of a surface. In this STM image of an orderly silicon crystal plane, the rows of atoms are 1.63 nanometers (billionths of a meter) apart, and the individual atoms within the rows are separated by a distance of 0.77 nanometer.

Atoms as Resonators

Perhaps no single technique or discovery more spectacularly illustrates the twentieth-century conquest of the atom than nuclear magnetic resonance (NMR)—a phenomenon that has transformed the process of chemical analysis and, in its medical incarnations, eliminated the need for thousands of painful and hazardous exploratory surgical operations every year.

In 1922, Otto Stern and Walter Gerlach conducted an experiment that unknowingly anticipated a basic property of subatomic particles. In their experiment, they showed that silver atoms behaved, in effect, like tiny bar magnets with a north and south pole. Three years later, in 1925, the critical insight came that particles have an intrinsic

Magnetic resonance was discovered by I. I. Rabi in this laboratory in 1938. The laboratory apparatus is a molecular beam system that allows atoms or molecules to be studied individually. The molecules moved from left to right through the evacuated tube shown in the center. Magnetic resonance was extended to bulk matter just after World War II by Edward M. Purcell and Felix Bloch (see next plate). Later, magnetic resonance became the basis for a powerful diagnostic tool for physicians, magnetic resonance imaging (MRI).

"spin." It was the spin of the outermost electron in the silver atom that gave the dramatic results observed by Stern and Gerlach. Now physicists know that systems of many charged particles, such as an atomic nucleus, have a net collective magnetic property called spin that is unique to each element.

In 1938, American physicist I. I. Rabi and colleagues found that when beams of molecules are placed in a strong external magnetic field, many of the nuclei try to align themselves with the outside field. But they do so incompletely, wobbling on their axes like a top that is slowing down. Rabi's group showed that, in that condition, if the nuclei are struck by a second, oscillating magnetic field from an electromagnetic wave that is at exactly the same frequency as their rate of wobble (the "resonant" frequency), they will absorb the field energy and "flip" their spin states—that is, reverse their north and south poles. This process diverts the molecules from the beam in a way that can be easily measured.

In the 1940s, Edward M. Purcell of Harvard and Felix Bloch of Stanford found other methods to induce and measure this effect. Soon NMR was being used to examine the composition of chemical compounds by detecting their resonances. Not only does each element have its own distinctive resonant frequency, but in compounds that frequency varies in slight but predictable ways as the magnetic fields of different kinds of neighboring atoms influence the target nuclei of the element in question—allowing NMR researchers to determine the structure of an unknown molecule.

Several scientists soon realized that NMR could also be used for medical applications. If living tissue were placed in a strong field, and some of its hydrogen atoms were spin-flipped, the resulting emissions could be assembled by software into an image of the relevant body part. By the 1980s, magnetic resonance imaging, or MRI, was in widespread use. And a seemingly arcane property—unveiled in the quest to understand the atom—has saved countless thousands of lives.

One of the most precise means of determining the structure of a material is by measuring the magnetic properties of the nuclei in its constituent atoms. Basically, this entails placing a sample in a strong magnetic field, which tends to align the sample's nuclear magnets parallel to the applied field. Some of the aligned nuclei will flip—that is, reverse their magnetic orientation—if exposed to an oscillating field of exactly the right "resonant" frequency, which depends on the particular nucleus and its local environment. So one way to determine the structure is to put various frequencies into the sample and see which ones are absorbed. This nuclear magnetic resonance cell, used by Harvard physicist Edward M. Purcell and colleagues Robert V. Pound and Henry Torrey in the 1940s, was filled with paraffin wax obtained at a convenience store.

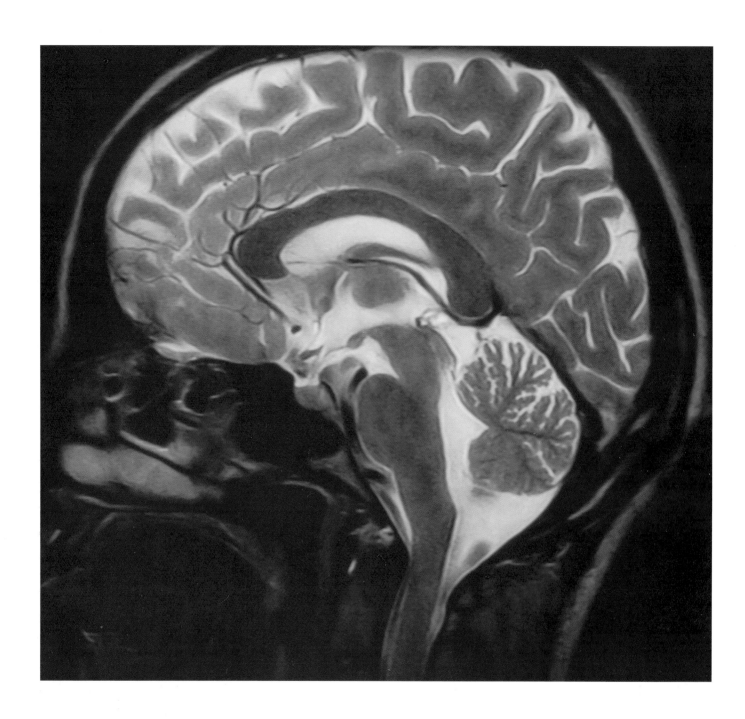

Unlike X-rays, MRI allows physicians to see and study soft tissue characteristics. Above, a magnetic resonance image through the middle of the human brain shows the different anatomical structures. MRI allows physicians to distinguish between the different tissues of the brain, such as the white/gray matter and the cerebrospinal fluid. Opposite is a magnetic image of the thorax, made while the patient is holding his/her breath. The image shows arterial vasculature including the heart, pulmonary arteries, and the aorta.

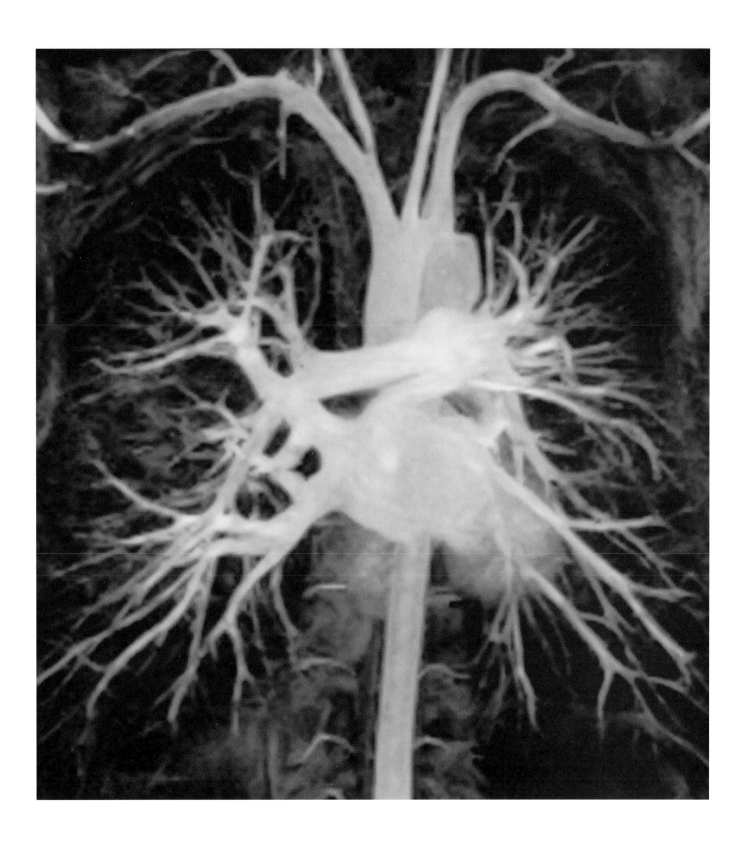

2 SPECTRUM

The last decade of the nineteenth century was positively aglow with scientific optimism. Amazing progress had been made in chemistry, astronomy, biology, and a dozen other fields. But no ensemble of achievements inspired more exuberant confidence in human ability to understand nature than the revolutionary discovery of how electricity and magnetism were interrelated, and how both were entwined in a wide spectrum of "electromagnetic" radiation—including light.

Isaac Newton's once authoritative conviction that light was made up of tiny particles had, it seemed, been completely discredited. In its place was a comprehensive wave theory, showing that types of radiation as seemingly different as warmth emanating from the sides of a wood stove, sunlight and candlelight, and radio signals were all variations on the same kind of thing. Each was made up of two components superimposed on one another: an oscillating electric field and a correspondingly fluctuating magnetic field. Electromagnetic waves could vary in length from a few trillionths of a meter to tens of thousands of meters.

This theory had been elegantly embodied in a set of equations devised by James Clerk Maxwell in 1873. By 1887, Heinrich Hertz had confirmed the model with dozens of convincing experiments demonstrating that electromagnetic radiation embodied many classic properties of waves, including interference, refraction, reflection, and polarization. Moreover, physicists had determined that light and all other forms of electromagnetic radiation traveled at only one speed, 300 million meters per second in a vacuum. (Speed is distance per unit time. For electromagnetic radiation, that means wavelength [distance] times frequency [number of waves per second]. Because the speed is constant for all kinds of radiation, those with longer wavelengths must have lower frequencies, and vice versa.)

Yet at the turn of the century, a few stubborn puzzles persisted. Their eventual solutions would reveal that light has a dual identity utterly different from what anyone imagined. And they would lead to a panoply of astonishing discoveries and inventions from photocells and radar to lasers and holograms.

Because of their extraordinary penetrating power, X-rays became a valuable diagnostic tool almost immediately after they were discovered by German physicist Wilhelm Röntgen in 1895. Shown here is one of his early X-ray images.

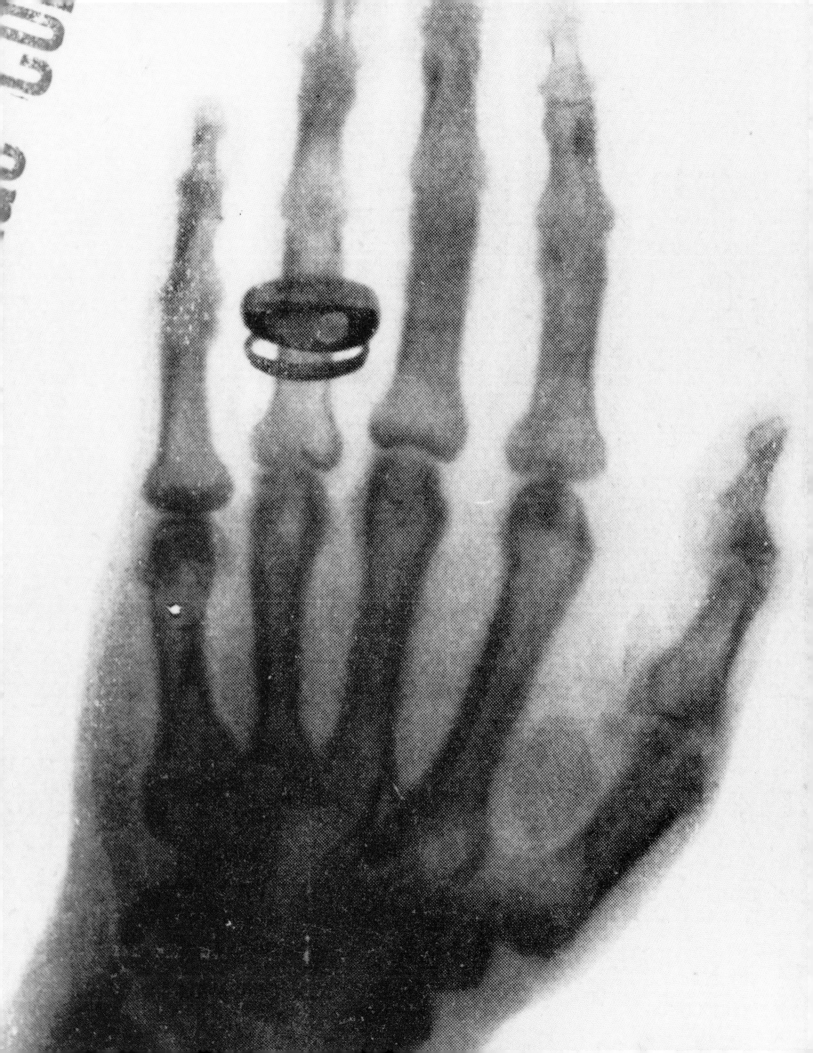

One of those puzzles was observed in the 1880s but never explained: when even a weak ray of light struck certain metals, electrons almost instantly flew off the surface in sufficient quantities to be detected as a small electrical current. This phenomenon came to be called the "photoelectric effect." Clearly the light was somehow ejecting electrons from the metal. But it did so in ways that couldn't be reconciled with orthodox wave behavior, which would have suggested that electrons would be ejected more slowly.

Furthermore, Philipp Lenard, one of Hertz's students, had shown unequivocally that increasing the brightness of light did not increase the maximum energy of the electrons emitted. That made absolutely no sense. An eight-foot-high wave crashing on a beach would convey vastly more energy to the grains of sand than a wave that was only six inches high. Even more puzzling, if the light was below a certain "threshold" frequency (which apparently was different for each kind of material), no electrons were ejected at all—no matter how bright the beam was! The photoelectric effect was an enigma.

The answer to this problem—which would revitalize Newton's particle theory of light and ultimately help transform the entire field of subatomic physics—came from a relatively obscure twenty-six-year-old German working in the Swiss Patent Office. His name was Albert Einstein. In 1905, Einstein postulated that light did not behave like a smoothly continuous wave. Instead, it propagated in discrete "packets," each of which had a specific energy content.

Five years earlier, German physicist Max Planck had proposed a related idea, namely that hot (that is, rapidly vibrating) matter gives off radiation only in individual units, or "quanta," and that the energy of each quantum is equal to the frequency of the vibration that produces it multiplied by a number now called Planck's constant.

Einstein argued that light was similarly quantized. The energy content of any individual bit was strictly a function of its frequency, as suggested by Planck's theory. (Twenty-one years later, American chemist Gilbert Lewis would give these individual chunks of light their name: photons.)

Einstein's idea explained why simply increasing the intensity of the incident beam on the metal surface would not raise the energy of the emitted electrons. It also made sense of the "threshold" effect: if the frequency (and hence energy) of the light was below the minimum needed to eject an electron, then there would be no effect on the metal, no matter how strong the light source.

Einstein's quantum theory—which obliged scientists to consider light as having both particle and wave properties simultaneously—was widely regarded as outlandish at the time. Planck himself didn't believe it. Even American physicist Robert Millikan, whose careful experiments between 1914 and 1916 would confirm the photoelectric effect and Einstein's predictions in superb detail, pronounced the idea of light quanta "wholly untenable." Most experts remained skeptical until 1922, when doubt suddenly became almost impossible.

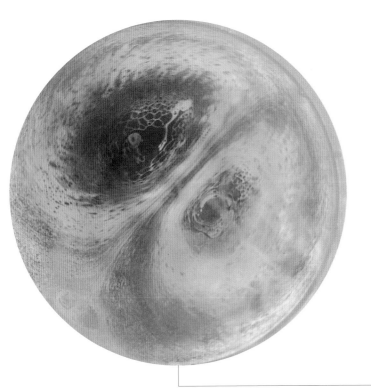

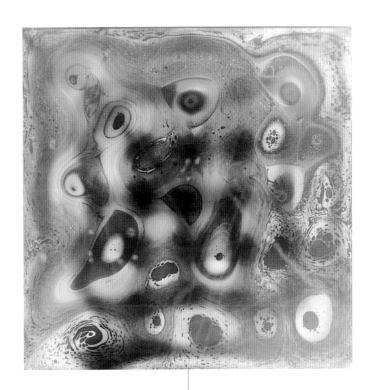

A defining characteristic of waves is interference. The wave nature of light is exhibited in these soap films, which are being vibrated. The vortex motion, flow patterns, and colors in the soap films are seen as interference fringes from the light source illuminating the films. The difference between the two patterns is caused by the shapes of the containers and the amplitudes and frequencies of the force driving the vibrations.

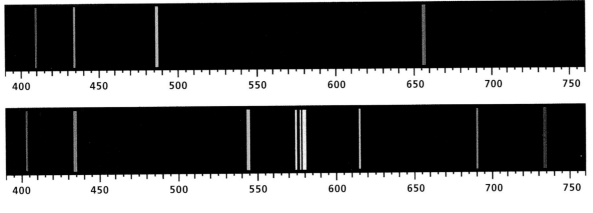

NANOMETERS

Just as each human being has a unique set of fingerprints, each chemical element has a unique arrangement of its component parts. As a result, each is capable of absorbing and emitting only certain wavelengths of light. When atoms of an element are heated or otherwise excited, the light they give off can be broken up into arrays of colored lines of different wavelengths, called spectra, in the same way that a prism separates white light into its constituent colors. Shown here are spectra for hydrogen (above) and mercury (below). In one of the earliest successes of quantum mechanics, Niels Bohr devised a theoretical model that predicted exactly the spectrum of hydrogen.

X-rays have been used in hundreds of ways during the century as physicists learned to direct and control them. In the photo above, dating from the turn of the century, X-ray equipment is used to diagnose injuries to a race horse in Lexington, Kentucky.

That was the year in which American physicist Arthur Compton discovered a seemingly inexplicable effect while working with "X-rays," which had been discovered by German physicist Wilhelm Röntgen in 1895 and so named because of their then unknown nature. Compton was bombarding blocks of graphite with X-rays and observing how the rays scattered in transit. To his surprise, he found that when the rays were deflected by a large amount, their wavelengths became as much as 7 percent longer. That is, their frequency became 7 percent lower.

There was no way for Compton to explain his peculiar findings using conventional notions of radiation. But they were thoroughly explicable with the Planck-Einstein theory. X-rays were behaving like individual solid particles. When an X-ray photon struck an electron, it would transfer some of its momentum to the electron in the collision. As a result, the X-ray's path would be bent and its energy content diminished.

Because electromagnetic energy is a function of frequency, the X-ray would emerge from the impact with a lower frequency and correspondingly longer wavelength.

Compton determined that the energy loss was directly proportional to the angle of deflection, just as it was for colliding balls on a pool table. Paradoxically, he wrote, the "wavelike property could be understood only in terms of particle-like behavior." Einstein was vindicated and the exotic idea of wave-particle duality was becoming comprehensible.

As is often the case, fresh concepts brought fresh problems. Apparently the photon had momentum. Yet the simple, Newtonian definition of momentum is mass times velocity. And the photon had no mass! As we shall see in a later chapter, the answer would emerge with the realization that, in the most fundamental sense, mass and energy are equivalent and interchangeable. The relationship was expressed by Einstein in the most famous equation in the history of physics: $E = mc^2$. So a photon's energy content lets it act as if it had mass.

The photoelectric effect would soon lead to a bevy of practical applications. The most familiar today are the photocell sensors that turn lights on at night or open doors automatically. Both are triggered by an interruption in the minute trickle of current produced when photons hit a receptive surface.

The phenomenon was also exploited in imaging techniques using the charge-coupled device, or CCD—the basis for TV cameras, home camcorders, and highly sensitive astronomical recording devices—and in the photovoltaic "solar panels" that provide energy for satellites, space probes, watches, and handheld calculators.

At the same time, the reverse of the photoelectric effect made possible the development of light-emitting diodes, or LEDs. Just as an incoming photon can kick an electron out of a crystal, physicists found, a photon is emitted when even a faint current is applied to an electron-depleted region in a crystal. The reason is that it is energetically costly to keep a negatively charged electron separate from a positively charged depleted spot, or "hole," to which it is attracted. When the two opposites combine, the energy that was once required to hold them apart is released as light. Hence the myriad solid-state lamps that wink like fireflies throughout modern life.

Meanwhile, the explosive growth in understanding photon behavior and the nature of radiation was providing science with unprecedented insight into the way atoms are arranged in materials.

Again, X-rays would prove fruitful both as an object of study and an experimental tool—once scientists were able to determine whether they were a form of electromagnetic radiation or something else entirely.

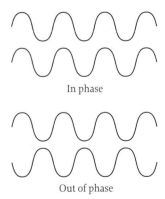

In phase

Out of phase

One definitive test would be to see if they exhibited a uniquely distinctive kind of behavior called interference—the process whereby waves reinforce one another or cancel each other out as they pass through a device that puts them in and out of phase with each other. In the early nineteenth century, British physician Thomas Young had shown that when visible light was made to pass through two adjacent slits in a barrier, the interaction of the wave fronts from each slit produced a typical alternating light-and-dark interference pattern. Since then, detecting interference had been the gold standard for demonstrating wave behavior.

But if X-rays—which could easily penetrate even opaque materials such as human limbs—*were* waves, they clearly had extremely short wavelengths (and correspondingly high frequencies and energies). And there were no artificial gratings or apertures small enough to provoke the kind of interference effects readily seen in visible light.

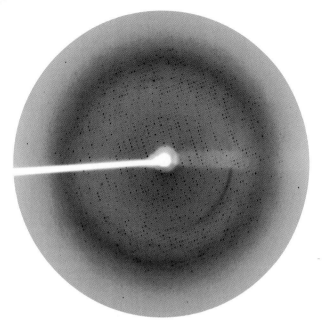

German physicist Max von Laue came up with a way around the problem, and in the process conclusively demonstrated the wave nature of the rays. If earlier estimates of the X-ray wavelength were approximately accurate at around 10^{-10} m, a distance we now call an Angstrom, then they were about the same size as the naturally occurring spacing between atoms in a crystal. So if two X-rays were to bounce off two adjacent planes of the crystal's highly regular layered arrays of atoms, von Laue reasoned, the difference in the distance each ray traveled might be just enough to produce a classic interference effect.

In 1912, von Laue's colleagues shot a stream of X-rays at a crystal of zinc sulfide and found that the beams exited in distinctive and wonderfully symmetrical geometric patterns. As would be expected from interference, the patterns differed according to the angle at which the X-rays struck the crystal. And, of course, different crystals generated different patterns.

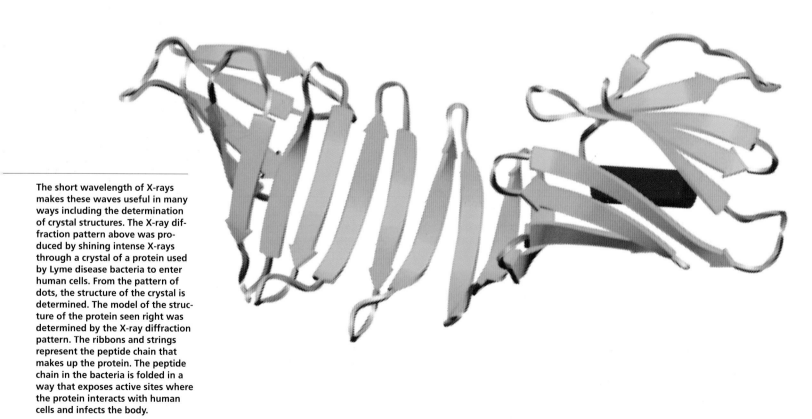

The short wavelength of X-rays makes these waves useful in many ways including the determination of crystal structures. The X-ray diffraction pattern above was produced by shining intense X-rays through a crystal of a protein used by Lyme disease bacteria to enter human cells. From the pattern of dots, the structure of the crystal is determined. The model of the structure of the protein seen right was determined by the X-ray diffraction pattern. The ribbons and strings represent the peptide chain that makes up the protein. The peptide chain in the bacteria is folded in a way that exposes active sites where the protein interacts with human cells and infects the body.

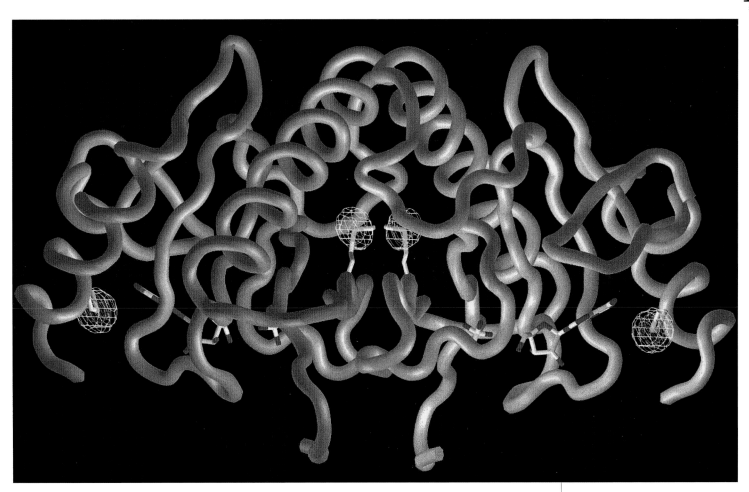

Shortly thereafter, X-rays became as powerful a tool of physics as they were of medicine. Just as von Laue had reasoned that a crystal could provide information about X-rays, two British physicists determined a few years later that X-rays could disclose secrets of crystal structure in unprecedented detail. After all, if the wavelength of the X-ray was known, then the resulting interference pattern of the emitted rays would reveal the spacing between planes of atoms.

The pair, William H. Bragg and his son William L., worked out a mathematical system for inferring precise crystalline structure from the angles observed in such images. The science of X-ray crystallography, as it became known, would evolve to uncover the configuration of scores of compounds crucial to physics, chemistry, and medicine. Several Nobel Prizes—including the 1964 award to British researcher Dorothy Hodgkin, who first determined the molecular structure of penicillin and vitamin B-12— have depended on this method. Another notable discovery made possible by X-ray diffraction is the dismayingly complex but eerily graceful architecture of intertwined spirals that would prove to be DNA—the molecular instruction set for creating life.

The Advanced Photon Source at Argonne National Laboratory is the most intense source of X-rays in the world. These intense X-ray beams now allow scientists to determine the structures of very complicated biological molecules. This image shows the three-dimensional structure of the protein, fragile histidine triad (FHIT), which is thought to be important for suppressing human cancer tumors. The X-ray diffraction technique used has incredibly high resolution (1.9 Angstrom) and allows measurements to be made at many different wavelengths. By looking for small changes between the results taken at different wavelengths, it is possible to determine the exact locations of key atoms in the molecule, in this case selenium atoms, whose position is shown by the small white cages in the image. Knowing the structure of such molecules will allow scientists to develop new therapies and treatments that would have been otherwise impossible.

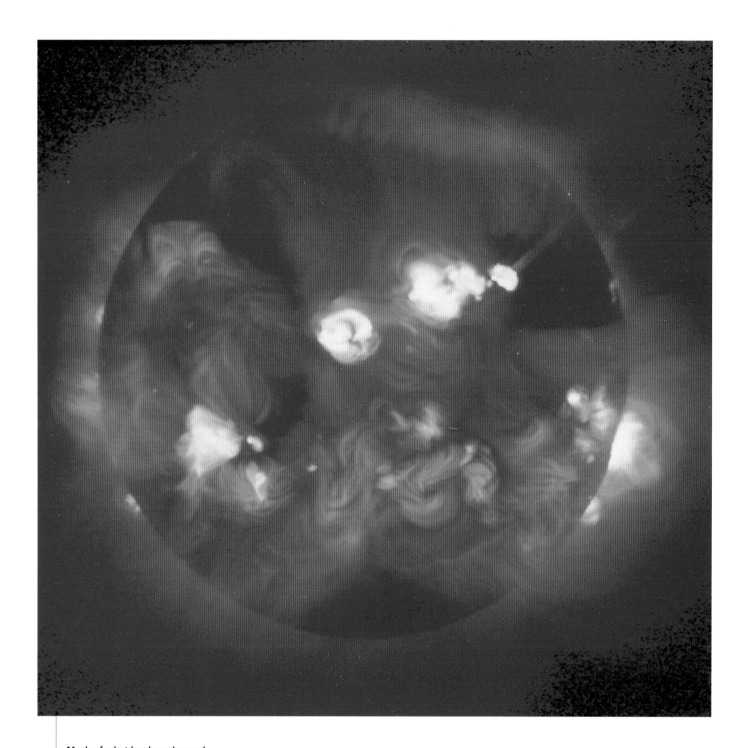

Much of what has been learned about the composition and structure of the cosmos in the past forty years has been revealed by examining data taken in wavelengths other than visible light. This X-ray image of the sun, taken on November 12, 1991, shows large differences in temperature and density of the solar plasma.

THE ELECTROMAGNETIC SPECTRUM

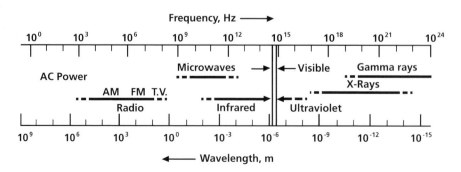

Frequency, Hz →

10^0 10^3 10^6 10^9 10^{12} 10^{15} 10^{18} 10^{21} 10^{24}

AC Power

Microwaves → ← Visible Gamma rays

AM FM T.V. X-Rays

Radio Infrared → Ultraviolet

10^9 10^6 10^3 10^0 10^{-3} 10^{-6} 10^{-9} 10^{-12} 10^{-15}

← Wavelength, m

When the torrent of charged particles that streams off the sun (called the solar wind) hits the Earth's magnetic field, many of the particles are trapped around the poles. When these particles collide with air molecules, they emit radiation that causes the "Northern Lights," or aurora borealis, and their southern counterpart. Other planets with magnetic fields—including the "gas giants" of the outer solar system—also have auroras, although they may occur at wavelengths that are not visible. Shown here is the first image of Saturn's ultraviolet aurora taken by the Hubble Space Telescope in October 1997. These curtains of light encircle both poles of the planet and rise more than 1,000 miles above the cloud tops.

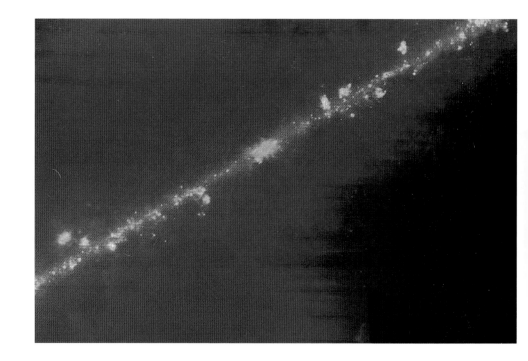

These images, incorporating six months of observation by the international Infrared Astronomical Satellite, show a bright horizontal band that is the plane of our Milky Way galaxy. Giant clouds of interstellar gas are seen as yellow and green knots and blobs in the enlarged section, right. The large yellow bulge near the middle of the band is the center of the Milky Way, approximately 30,000 light years from Earth.

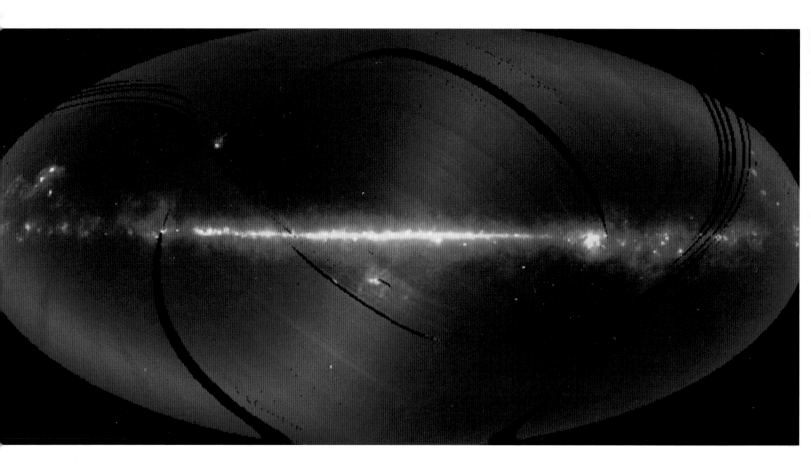

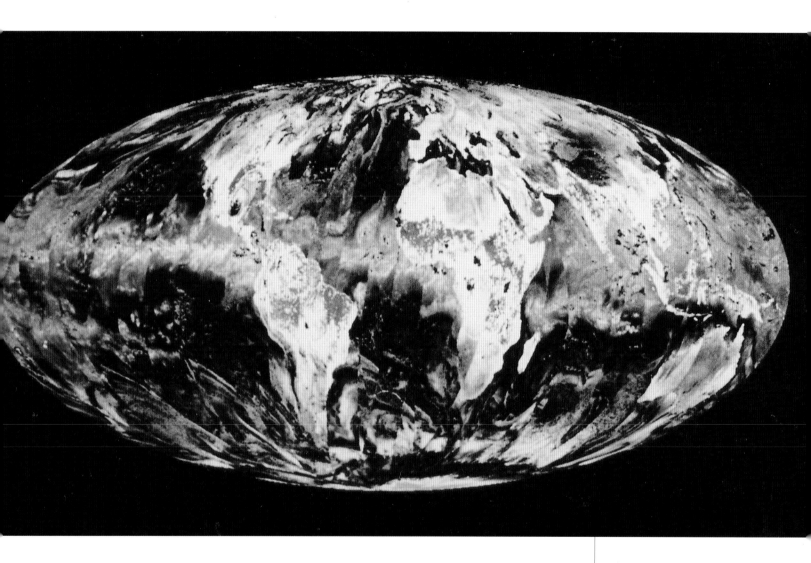

The Earth emits radiation of various wavelengths. This computer-reconstructed image shows the planet in microwaves, as detected by an orbiting satellite over four days in 1988. Microwave emissions reveal details of air conditions and surface characteristics that cannot be observed in visible light.

Making Waves

While hundreds of labs were at work on high-energy radiation, even more were learning to control various frequencies below visible light, including those that would eventually carry radio, television, and radar signals.

The basic principles necessary for generating all three had been established in the 1880s by Hertz, who succeeded in transmitting the invisible waves across his lab. A moving charge (whether positive or negative) creates an electromagnetic wave. So if electrical currents of various frequencies flow back and forth through an antenna, they produce waves of the same frequencies that are emitted from the antenna. The waves propagate outward at the speed of light; when they encounter a receiving antenna, or indeed any metal object in their path, they cause electrons within the metal to start moving back and forth at the same frequency as the waves. Thus the initial current pattern, or signal, can be re-created in objects hundreds of miles away. By 1901 Italian physicist Guglielmo Marconi, after exhaustive experiments with different kinds of antennas, had been able to send a simple "wireless" Morse code signal over the Atlantic. Civilization was on the brink of truly instantaneous global communications. But there was one forbidding obstacle.

Radio signals were often appallingly faint by the time they reached their destination, and there was no suitably sensitive but powerful way to boost them at the receiving end. Receivers could barely discern the gross on-and-off pulses of Morse code. Detecting and amplifying the subtle frequency differences in human speech—although theoretically quite simple—was in fact nearly impossible.

A glimpse of the solution had come in the early 1900s when British researcher John Ambrose Fleming created a device he called a diode. It relied on a well-known effect observed by Thomas Edison and others (and very similar to the cathode ray phenomenon studied by Thomson): when a negatively charged electrode, or cathode, is heated, it emits electrons that are drawn to a second, positively charged electrode called an anode. Fleming found that if he fed an alternating current—that is, one that oscillated between positive and negative charges—into the anode, the positive peak of each wave produced an attractive force strong enough to cause electrons from the cathode to jump across to the anode. The greater the positive charge, the more electrons moved across the gap.

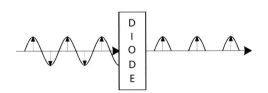

Because the electrons jumped only when the charge on the electrode was positive, the "thermionic" diode served as a dandy rectifier—that is, a device for converting alternating into direct current. More important was the basic principle: a diode was capable of detecting even fairly subtle changes in a variable current. And a variable current, weakened by distance, was precisely what radio waves induced in antennas. Communications had become "electronic."

As a sensor, the diode was fairly crude. But in 1906, American engineer Lee De Forest invented a critical variation in the design. The result was the vacuum tube that powered electronics for half a century.

His device was a "triode" composed of a heated cathode, an anode, and a novel third electrode in the form of a grid placed between the other two. The grid carried the weak fluctuating radio signal, or whatever needed to be amplified. As electrons streamed off the hot source and headed for the positive anode, they were accelerated or retarded by the grid depending on whether its charge was positive or negative at the time they passed. Thus even minor changes in the variable grid field, such as might be caused by radio waves, would have a large effect on the electron currents that flowed to the high-voltage anode, amplifying the pattern of those changes. By 1910, De Forest was broadcasting the voice of fabled tenor Enrico Caruso from New York's Metropolitan Opera.

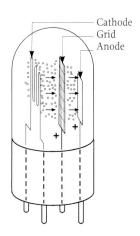

Cathode
Grid
Anode

There would be scores of subsequent innovations in the ability to send and detect radio (and soon TV) signals. But the fundamental physics would not change much until 1947, when the invention of the transistor (as we shall see in a later chapter) set off a second communications revolution that continues today.

Reflections on War

At the same time, scientists were investigating another property of radio waves: when they struck a solid object, they were reflected back toward their source. In 1904, German researcher Christian Hülsmeyer had patented a system based on that principle to prevent collisions between fog-shrouded boats and trains. Because the speed of light was a well-known constant, the time it took the reflections to return would reveal the distance of the object. The effect might also be exploited to detect remote entities such as warships, a use confirmed in 1922 when physicists at the U.S. Naval Research Laboratory in Washington, D.C., monitored a vessel moving on a river.

At that time, there was little apparent practical need for such devices—though they were proving their worth in geophysical science. Two separate teams had predicted in 1902 that the atmosphere contained an upper layer that is electrified (ionized by particles streaming off the sun) and thus reflects or refracts radio waves. This "ionosphere" makes it possible for electromagnetic transmissions to travel over the horizon, well beyond the line of sight. In 1924, a British group headed by Edward Appleton bounced a series of continuous waves off the upper atmosphere and observed telltale interference patterns that revealed the previously mysterious layer. The next year American physicists Gregory Breit and Merle Tuve used a pulsed beam to confirm the discovery and to reveal the height of the ionosphere at about 60 miles above the Earth.

But as war became menacingly likely in the 1930s, many nations began studying military uses of Radio Detection and Ranging, or radar. The premier problem was to make waves that were long enough to travel through the atmosphere without scattering, as visible light does, but short enough to give a reasonably detailed image of an object such as a ship or large airplane. (The same principle makes it impossible to observe molecules with a visible-light microscope: you can't see an object that is smaller than the waves used to observe it.) So ideally the radar waves should be no longer than a few yards at the most.

That was good news for antenna builders. A radar transmitter that generated waves a few yards long could be easily mounted on a ship or truck. But it was very bad news for physicists. Because the speed of light is a constant, the shorter you make the wavelength, the higher you have to make the frequency. Consequently, the frequency of useful radar waves would have to be hundreds of millions of cycles per second!

During World War II, radar was developed and improved at the Radiation Laboratory, familiarly known as the "Rad Lab," at the Massachusetts Institute of Technology. Pictured here is one of the research areas in November 1940.

That was a stupendous challenge. Electromagnetic waves, we have seen, are made by back-and-forth motion of electrical currents. To attain the needed frequencies, scientists would have to find a way to move charges faster than anyone had ever been able to do for radio waves. The challenge was met in 1939 by British physicists John Randall and Henry Boot who created the first practical magnetron, a device that produced waves by causing electrical currents to oscillate back and forth in a little cavity just as sound waves resonate in an organ pipe.

Numerous innovations followed rapidly. For example, by bouncing radar waves off a target and then analyzing the reflections for the Doppler effect (which shortens the wavelength of emissions from a source moving toward the detector, and lengthens those from a source that is moving away), observers could determine not only the position of a ship or plane, but the direction in which it was moving. The technology became a major factor in the Allied victory in World War II.

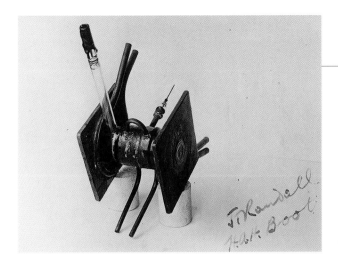

The development of radar technology was dramatically accelerated during World War II. A key event was the invention of the magnetron, a device that creates constant-frequency microwaves as charges bounce back and forth in a confined structure. Shown here is the first cavity magnetron built by British scientists J. T. Randall and H.A.H. Boot.

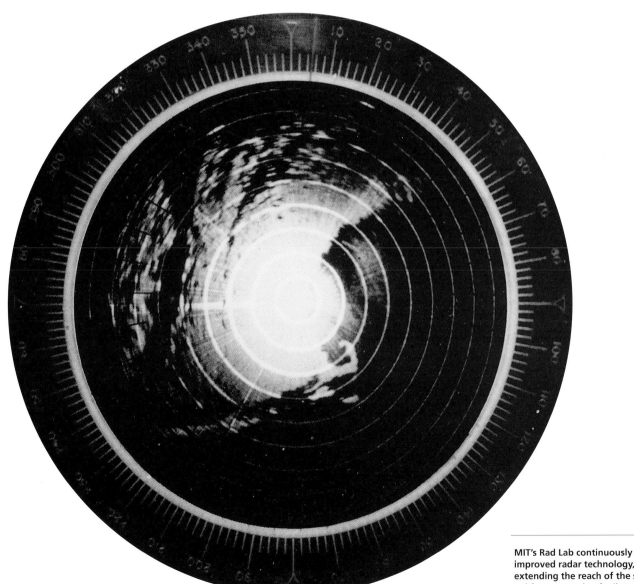

MIT's Rad Lab continuously improved radar technology, extending the reach of the signal from a couple of miles to hundreds of miles. Shown here is a radar image of New England, with the characteristic hook shape of Cape Cod clearly visible.

As scientific understanding of electromagnetic radiation grew during the twentieth century, and the technology to detect minute variations in wavelength and frequency improved, a host of practical devices were developed. Among the most exotic is the atomic clock, which keeps time by measuring emissions from atoms. Shown at right is the first working model of a cesium clock, built by Jerrold Zacharias at the Massachusetts Institute of Technology in 1953. The latest versions of atomic clocks—such as the one (opposite) from the U.S. National Institute of Standards and Technology—are used as master timekeepers to set other clocks and coordinate "universal time" around the planet. Clocks of this sort are accurate to one second in ten million years.

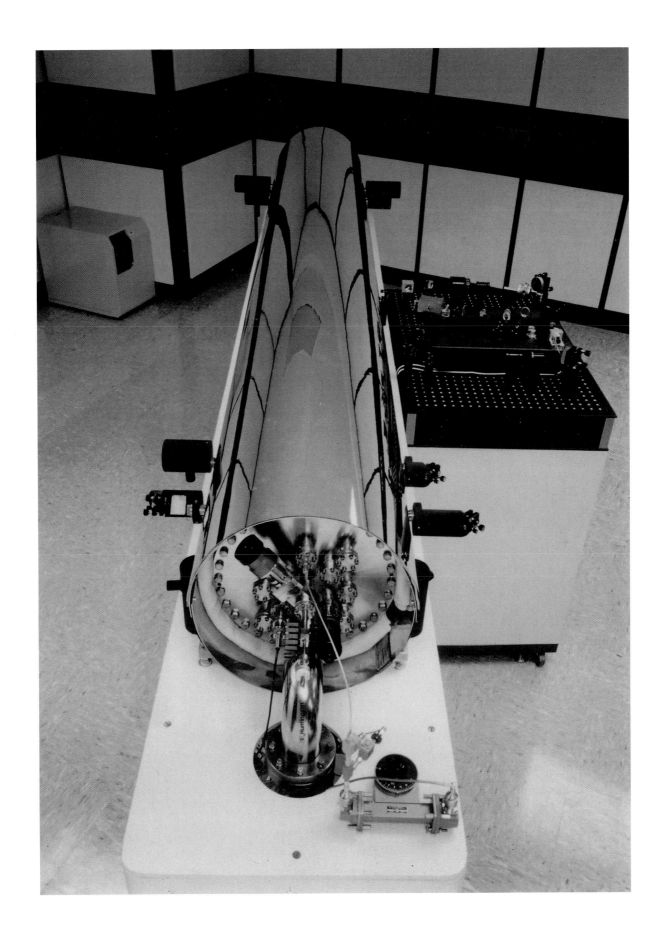

Latter-day improvements in the magnetron would eventually generate waves so short that they could shake individual water molecules. Consequently, the ordinary civilian is now able to use the once super-secret military technology of radar to bake potatoes or pot roasts in a microwave oven. It works because a water molecule, containing two hydrogen atoms and one oxygen atom, is electrically polar. The hydrogen atoms, instead of arranging themselves on opposite sides of the oxygen atom, both bind to one side, giving that side a net positive charge. The other has a net negative charge. So when water is exposed to an electric field (such as the one carried by an electromagnetic wave), its randomly oriented molecules begin to rotate in order to align themselves with the field. Experiments showed that microwaves with a frequency of around 2.45 billion cycles per second (gigahertz, or GHz) cause water molecules to twist so violently that they banged into their neighbors, causing vibrations and thus thermal energy. The heat generated cooks the food.

Improvements in detection have made it possible to obtain high-resolution pictures of the way water droplets or dust particles are moving in the air, giving weather forecasters a much more accurate method of tracking moving storms. And microwave beams at about 1.5 GHz from orbiting broadcast satellites are now used to produce the Global Positioning System (GPS). A ground-based GPS detector compares the different times it takes for signals to arrive from several different satellites of known locations and calculates its position to within a few meters.

In order to triangulate a position, the detector needs to know the precise time at which various signals left each of the satellites. In theory, because electromagnetic waves travel at 300,000,000 meters per second, getting a reading that is accurate to within, say, 10 meters requires a clock that can tell time to within at least 30 million parts per second. In practice, much more precision is needed, and GPS transmitters divide each second into slices that are 400 times smaller than that.

This is possible because every satellite carries an atomic clock. All clocks rely on something that oscillates at a constant frequency, whether it is a swinging pendulum or vibrating quartz crystal; the accuracy depends on the regularity of the oscillator. And physicists have long known from studying spectra that when atoms are energized with light or heat, they give off radiation at particular frequencies unique to each element and determined by that element's specific arrangement of electrons.

Atomic clocks take advantage of that phenomenon. Atoms of a certain kind, such as cesium, are electromagnetically excited, and the characteristic frequency of the radiation they absorb or emit is used as the "pendulum" of the atomic clock. The latest models exploit hydrogen atoms and have an accuracy of about 1 part in 10^{15}. That is equivalent to the gain or loss of no more than a single second in 30 million years!

Coherent Light

The emission effect that makes atomic hydrogen clocks so accurate is the same one that has led to the most spectacular embodiment of optical physics in the twentieth century: the laser. (The term is an acronym for light amplification by stimulated emission of radiation.)

Einstein had predicted the phenomenon as early as 1916, when he theorized that there were two ways for an atom to emit photons: spontaneous and stimulated emission. The first is the familiar type that occurs every day in luminous objects from the sun to lightbulbs. If an atom is excited by a quantum of energy that exactly corresponds to one of its specific allowed energy states (as Niels Bohr postulated), it absorbs the energy. But it is in the nature of atoms to seek the lowest energy level possible. So the atom sheds its excess energy almost immediately—typically within a nanosecond, or billionth of a second—by emitting a photon.

Stimulated emission is a kind of variation on this effect. Einstein proposed that if an electron was already in an excited state and was then struck by a photon of exactly the right energy, it would emit two photons: the original and a second photon identical to the first. Not only would the process double the number of photons, but each quantum of radiation would be perfectly matched, or "coherent." That is, the wave of the second photon would be exactly in phase with the one that stimulated its emission. Those photons would in turn strike other excited atoms, which would emit yet more stimulated radiation, amplifying the wave beam in a chain reaction.

By 1920, the theoretical basis for that effect—which was then called "negative absorption" of photons—was well laid out. Yet it took physicists nearly forty years to devise the first laser-like device. In part, that was because of traditionally sluggish communication between physicists working on fundamental problems and those working on applications (until World War II made them constant partners). And in part it was due to the fact that there seemed to be no practical use for a gizmo that, most experts assumed, was unlikely to produce more than a weak beam.

There were also two difficult scientific issues. One was that stimulated emission will only work if a majority of atoms in a given substance are in the excited state; otherwise, lower-energy atoms will absorb more photons than are emitted by high-energy atoms, and no amplification can occur. Normally, however, only a minority of atoms in a substance is in an excited state. The majority-excitement condition—called a "population inversion"—almost never happens in nature.

The other problem is that not every excited energy condition is suitable for stimulated emission. There are numerous competing processes, occurring simultaneously, that can affect the atoms' excited state, and a satisfactory device would have to eliminate or minimize them.

Finally, in 1954, American physicist Charles H. Townes created a device that satisfied both criteria. He segregated a population of excited ammonia molecules, bombarded them with microwaves, and proved that the microwave output was measurably stronger than the input. This maser (for microwave amplification by stimulated emission of radiation) produced a continuous microwave output so regular that it was used as the frequency standard for the first generation of atomic clocks.

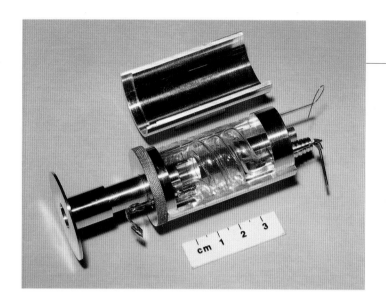

This is the original ruby laser, built by Theodore Maiman at Hughes Research Laboratories in May 1960. Barely five inches long, it produced laser beams by energizing a cylindrical ruby crystal (aluminum oxide) by means of a flash lamp coiled around the cylinder. The end of the cylinder on the right is covered with a mirror so that laser-generated photons traveling that way are reflected back into the crystal. The left side is half-mirrored, so some of the photons bounce back into the ruby and some escape down the barrel to the left, making up the laser beam.

The maser, a device that produces "coherent" microwaves (that is, waves that are exactly in phase with each other) was the precursor of the laser, which emits coherent visible light. In this photo, maser inventor Charles H. Townes is shown with the second version of his apparatus at Columbia University's Radiation Laboratory in 1954.

The latest laser systems are used to study both light and light's effects on various materials. Typically the equipment is secured to a massive, stationary platform called an optical bench in order to minimize low-frequency building vibrations during experiments.

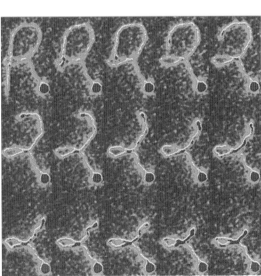

Laser beams can be used to manipulate microscopic objects by deploying them as "optical tweezers." One or more beams gently prod their objectives into position without damaging the material and without visually obscuring the effects that scientists want to study. Here physicists at Stanford University used tightly focused infrared laser rays to stretch out tiny coiled strands of DNA, each about 50 microns (millionths of a meter) long. They chemically doped one end of the DNA strand so that it would bind to a 1-micron bead of plastic, and then applied the optical tweezers to the bead. Using the bead as a "handle," they were able to tug the tangled DNA molecule into a variety of configurations, as seen in this composite image, and then examine the way it relaxed back into its initial shape. These and related experiments were used to confirm a leading theory of the way complex molecules called polymers are able to move.

One of the most exciting applications of laser light is its use in micromachining various materials by focusing the energy of many coherent photons into a very tight beam. In the photo above, laser light was used to cut notches in a single human hair less than 1/10,000 of an inch in diameter.

At about the same time, Russian physicists Nikolai G. Basov and Alexander M. Prokhorov had independently proposed similar devices, and those plans led eventually to the creation of semiconductor lasers. Harvard physicist Nicholas Bloembergen showed how solid-state masers could be fine-tuned, making them important research tools for atomic spectroscopy. They have since been used repeatedly when the greatest sensitivity is needed, as in radio astronomy.

Shortly after the first maser demonstration, Townes and Arthur Schawlow at Bell Labs came up with a plan to make masers that worked at the much shorter wavelengths at or near the visible part of the electromagnetic spectrum. (This was no trivial task. Townes's maser microwaves were in the centimeter range; visible wavelengths are nearly 100,000 times smaller.) Using those ideas, American physicist Theodore H. Maiman created the first working laser in 1960 with a design that is still in widespread use today.

The device employed a cylinder of ruby, a compound of aluminum and oxygen with imbedded chromium impurities that produce the crystal's characteristic red color. In order to boost a majority of chromium atoms into an excited state, Maiman wrapped a flash lamp (in this case, a coil of glass tubing filled with xenon gas) around the ruby cylinder. Every time a pulse of current was run through the xenon, it emitted photons that "pumped" the chromium into a population inversion.

One end of the ruby cylinder was completely covered with a mirror, so that pairs of photons striking it were reflected straight back into the crystal, where they would stimulate further emissions. The other end was only half silvered, allowing a burst of coherent photons to emerge from it every time the lamp flashed.

Several alternative designs soon followed. Other Bell Labs researchers devised the first gas laser, using an electrical discharge through a mixture of helium and neon. Although originally derided by some as "useless," lasers were rapidly improved, and four decades later, they are an indispensible tool of modern civilization.

Lasers now provide the data streams for worldwide fiber-optic communications lines. At high energies, the photon beam makes materials hot enough to break atomic bonds, and lasers are used for a number of cutting and welding operations in industry. At lower energies, they are employed to perform various kinds of nearly bloodless surgery (notably on eyes), to control computer printers, and to measure the distance between Earth and the moon to an accuracy within 50 feet. Lasers are in every bar-code reader at grocery counters and drug stores, and they extract the music or software information from every compact disk. They make possible three-dimensional holograms, perfectly straight lines for surveyors, enormously precise measurements of small distances, and scores of analytical techniques. New uses are emerging daily.

In the course of the century, physics had completely transformed human understanding of light. But as it turned out, the illumination of the photon was only one of the shocking surprises lurking in the strange new world of the quantum.

The stupendous demands of modern communications have prompted physicists to look for ways to compress more information into less space. One of the most successful is the use of light instead of electric currents to carry data. The frequency of visible light (hundreds of trillions of cycles per second) is about a billion times greater than the basic operating frequency of a high-tech desktop computer (around 100 million Hz), giving light a far greater information capacity per unit of length or time. That is why 90 percent of intercontinental telephone calls are now carried on fiber-optic cable. According to one estimate, approximately 400,000 miles of such cables will cross the oceans by the year 2005. The challenge in creating these high-volume conduits was to devise fibers that would carry photons with minimal loss. That required nearly perfect reflection from the fiber walls, as well as extremely uniform optical properties to avoid altering or attenuating the signal as it moves along the fiber. Today, a single hair-thin fiber can transmit several hundreds of thousands of telephone conversations simultaneously. Here light enters one end of a glass fiber, is transmitted by the fiber around and around the spool, and finally exits from the opposite end of the fiber.

3 QUANTUM ⁶⁴

Of all the discoveries that would shake science to its deepest foundations during the twentieth century, none matched the impact of quantum mechanics—a revolutionary way of looking at nature on the smallest scales. It postulated radical new concepts that even Einstein, who helped start the revolution, soon found completely intolerable. It made outrageous predictions, some of which were eventually confirmed to a precision of one part in 10 billion, making quantum mechanics the most accurate theory in the history of science. And it would lead to development of a host of practical devices including transistors, enormously accurate clocks, and super-sensitive detectors that have changed—and still are changing—the lives of all of us.

But most important, it shattered many of humanity's most firmly established convictions about matter and energy. In Newton's congenial cosmos, every entity of whatever size was presumed to have a set of definite, unambiguous properties at any moment in time. Thus if one could somehow know all the properties of all the components of a given system, one could predict precisely how it would behave in the future or how it had changed in the past.

That method had worked so well that in 1846 astronomers had discovered the planet Neptune by inferring its existence from observed irregularities in the orbit of Uranus. Using Newton's laws of motion and gravitation, they calculated what the path of a massive object would have to be in order to perturb Uranus' motion in that way. And when they aimed their telescopes, they found Neptune within 1 degree of its predicted position!

But in the atomic realm, quantum mechanics utterly rejected that kind of hugely successful, deterministic thinking. In those dimensions, physicists found to their amazement, the specific characteristics of particles, or even whole atoms and molecules, were inherently unknowable until the instant they were measured.

Moreover, certain kinds of measurements—such as the simultaneous determination of momentum and position—were intrinsically impossible. Even more baffling, objects no longer fit into comprehensible categories. Light was not exactly a wave; an electron was not altogether a particle. Each was somehow both a wave and a particle *at the same time* and could act as either one or the other as circumstances required.

This photograph shows the blackboard in Richard Feynman's office at the California Institute of Technology. In the upper left-hand corner, he wrote: "What I cannot create, I do not understand."

hat sense is what happens | Describe
plane in a string independent | center of
hat happens at another | string
t later in the string but distant? | without
| the rules

$$P(S_a = +, S_b = + | "\delta", \hat{a}, \hat{b}) = \int$$

$$= \frac{5}{B}$$

$$d\hat{r} \equiv d\frac{\Omega}{4\pi}$$

$$|\hat{3}\rangle$$

$$\hat{a}$$

$$\frac{1 + \vec{a} \cdot \hat{3}}{2} = \int d\hat{r} \, u(\hat{r}, \hat{a})$$

In 1927, Clinton Davisson (left) and Lester Germer provided convincing evidence for one of the strangest notions in the history of science, namely that moving particles behave like waves, with a wavelength that depends on their energy. One distinctive property of waves is diffraction, the bending of waves when they encounter an obstacle or aperture. Davisson and Germer showed that this happens to electrons striking a piece of metal. The effect is seen clearly if the electrons have an energy that makes their wavelengths similar to the spacing between layers of atoms in the metal crystal.

The apparatus used by Davisson and Germer in their New York laboratory was a glass vacuum tube containing an electrode that fired electrons at a small block of nickel. The physicists measured the angle at which the electrons were scattered from the metal by using the semicircular compass at the center of the tube. At first, the electrons scattered in all directions, as would be expected if electrons behaved only as particles. But when an accidental contamination of the block forced the scientists to bake impurities out of the metal at high temperature, the nickel atoms rearranged themselves from a random configuration into a very regular crystal lattice. Suddenly the electrons began emerging in particular patterns revealing their wavelike nature.

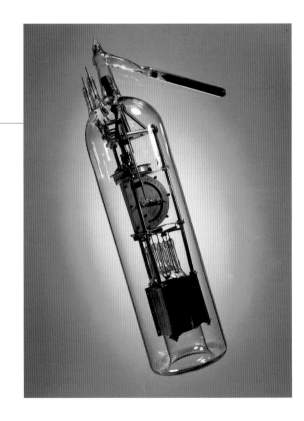

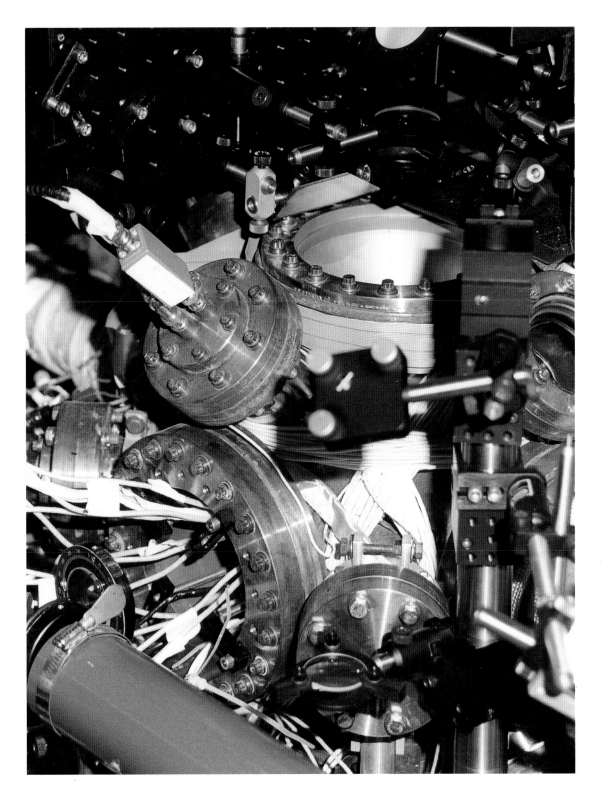

Among the exotic predictions of quantum physics is one made by Einstein and Indian physicist Satyendra Nath Bose in the 1920s. Because of the wave nature of matter and the fact that objects of lower energy have longer wavelengths, a group of neighboring atoms should behave in a very peculiar way when made extremely cold. At the lowest energy levels, their wavelengths should actually overlap until the entire array behaves, in effect, as one agglomerated superatom, or "condensate," as Bose and Einstein called the then hypothetical condition. This 1997 photo shows experimental apparatus used to observe a Bose-Einstein condensate at the Massachusetts Institute of Technology. Atoms are trapped and chilled in a vacuum chamber, and then photographed with cameras pointed through the port at the top of the picture.

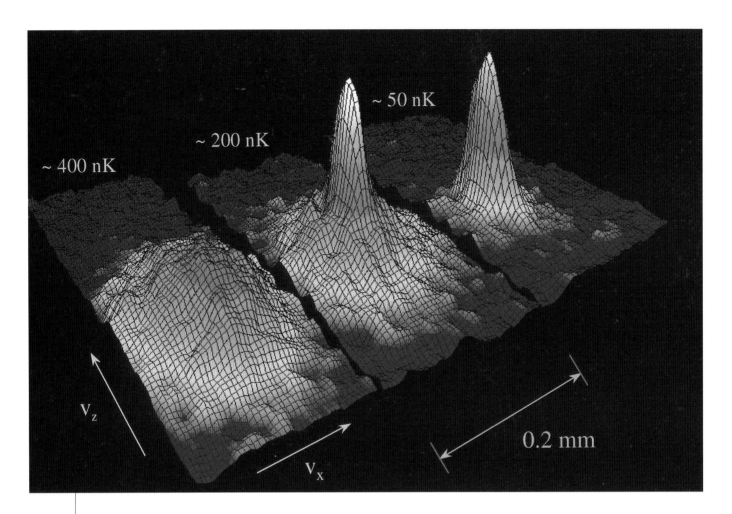

~ 400 nK

~ 200 nK

~ 50 nK

V_z

V_x

0.2 mm

The first Bose-Einstein condensate was produced on July 14, 1995, when experimenters in Colorado were able to trap rubidium atoms at the required temperature, which is only a few billionths of a degree above absolute zero. These computer-generated images show how, as the temperature drops (from left to right), more and more of the trapped atoms occupy the same quantum state, represented by the height of the spike in the center of the grid.

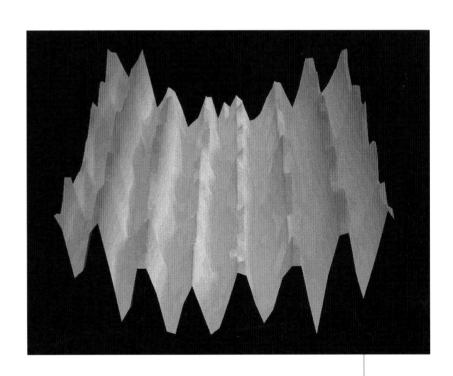

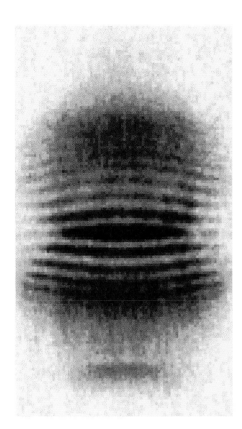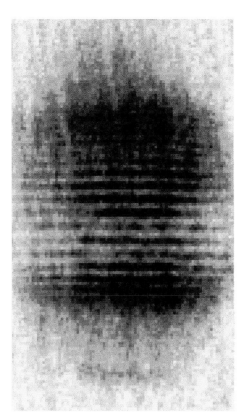

Unlike many explosive changes in scientific thought, the amalgam of concepts collectively called quantum mechanics was not the product of one or even a few individuals. It emerged gradually during the first quarter of the century in the work of more than a dozen physicists in a handful of countries.

As we have seen, the first tremors began in 1900 when Planck determined that the atoms of heated bodies radiate energy only in specific, discrete quantities. Then in 1905, Einstein argued that light is quantized—that is, made up of separate units or photons that have different energies depending on their frequency, a notion akin to Newton's belief that light was a stream of particles. This insight, however, did not displace the amply confirmed idea that light was a wave. Instead, it showed that electromagnetic radiation has both wave and particle properties simultaneously. To those trained in classical mechanics, that inherent duality was extremely hard to swallow.

But it was far easier to take than the analogous idea proposed in the early 1920s by French physicist Louis de Broglie to explain why atoms emit light only in specific quanta. He theorized that, in some way unknown to orthodox physics, the electrons encircling an atomic nucleus are behaving like waves, with regular, oscillating patterns similar to mechanical or acoustic waves in musical instruments. In addition, he devised a formula that described the wavelength of an electron moving at a particular speed.

By the logic of wave behavior, two Bose-Einstein condensates that are brought into contact with each other should show interference effects. Just as two water or light waves coming together reinforce each other when a crest adds to a crest or cancel each other out when a crest encounters a trough, the matter waves of two condensates should interfere with each other producing a pattern of alternating positive and negative reinforcement. In the images above, two expanding B-E condensates—each consisting of approximately 50,000 sodium atoms—released from a trap merge into each other and show interference fringes. The two pictures show these fringes in the condensates for different degrees of their overlap. The interference fringes demonstrate that B-E condensates are made up of atoms occupying a single quantum state and that the atoms are wavelike. The computer-generated image, opposite bottom, is another way of showing the interference.

Danish theorist Niels Bohr played a major role in the conceptual revolution that led to our modern understanding of atoms. In 1913, he published a seminal paper proposing a structure for the hydrogen atom. Bohr's scheme combined some quasi-Newtonian ideas with elements of the new "quantum" physics emerging in the first quarter of the twentieth century. It would later be substantially revised. But it predicted the spectral emissions of hydrogen with excellent accuracy.

German mathematical physicist Werner Heisenberg created the first version of the new physical theory of quantum mechanics at the age of twenty-four. Two years later, in 1927, he published his famous "uncertainty principle"—a concept that effectively brought to completion the formal structure of quantum mechanics.

Austrian theorist Erwin Schrödinger created his own version of quantum mechanics in January 1926. It was based on more familiar mathematical methods than the analysis proposed by Heisenberg, and was more easily accepted by physicists. In an amazing outpouring of creativity, Schrödinger published a major paper every month during the first half of 1926. One of those demonstrated that his model and Heisenberg's were equivalent.

Many of the great architects of quantum theory saw one another frequently, often at symposia in Europe, until World War II temporarily curtailed free association. Here Heisenberg (left) and Bohr are pictured at lunch around 1930.

English theoretician Paul Dirac began making crucial contributions to quantum physics at the age of twenty-three. In 1928, he published a full quantum treatment of the electron. Within that analysis was the suggestion that the electron had a mysterious opposite, or "antiparticle," which is now known as the positron. It would be discovered four years later.

A plucked guitar string, for example, can vibrate at one lowest, "fundamental" frequency or at various higher frequencies called harmonics whose wavelengths are simple fractions of the string's length. At any other wavelengths, the vibrations will cancel each other out and will not resonate.

Analogously, de Broglie reasoned, an electron can occupy only certain orbits. One is its lowest-energy "ground" orbit—the atomic equivalent of the guitar string's fundamental frequency. The other allowed orbits are those whose circumferences are a multiple of the electron's wavelength.

That property, he showed, could explain why electrons do not inhabit an infinitude of arbitrary states, but only the specific permitted orbits described by Niels Bohr in his original model of atomic structure. Each permitted orbit above the ground state can be occupied only by electrons with specific additional amounts of energy. (Similarly, the higher your floor in an office building, the more elevator energy it takes to get you there, and the harder you will hit the ground if you fall out the window.) And electrons can only move between the allowed orbits. (Much as an elevator will only stop at specific floors and not five-sixths of the way between two floors.)

That, in turn, explains why each atom absorbs and emits a distinctive spectrum of photons. If an incoming photon has exactly the amount of energy required to raise an electron from one orbit to another, then the atom will absorb that photon. And when an electron drops from a higher permitted orbit to a lower one, it gives off a photon whose energy content is precisely the same as the energy difference between the two orbits.

In fact, de Broglie's theory demands that *all* moving objects—not just electrons, but baseballs, birds, or Buicks—have an associated wavelength. But we can't observe those effects in the visible world because the wavelength gets smaller as the mass and velocity get larger. Thus a bowling ball rolling down the alley has a wavelength much shorter than any existing device can measure.

At first, this notion was widely regarded as preposterous. One wag dismissed it as "La Comédie Française." After all, unlike the equally novel theories of Planck and Einstein, the wave nature of matter was mere conjecture, not based on any direct experimental evidence.

Yet within a couple of years, to the astonishment of skeptics, the weird effect had been confirmed in two independent labs. Just as researchers investigating the nature of X-rays had looked for interference phenomena to test whether the rays were really electromagnetic waves, the quantum experimenters assumed that if electrons actually had wave properties, they would reinforce or cancel each other out in the right circumstances. And the right circumstances were available: De Broglie's equations predicted that the wavelength of a moderately energetic electron (about 100 electron volts) would be around a billionth of a meter, conveniently about the same as the spacing between planes of atoms in a metallic crystal.

Meanwhile, American physicists Clinton Davisson and Lester Germer had been bouncing a beam of electrons off a nickel surface to study the crystal structure. The results puzzled them. Attending a scientific meeting in 1926, Davisson was amazed to hear other physicists suggest that the experiment had detected electron wave properties. Back in his lab, he and Germer improved their apparatus to produce a clear result. The electron waves were either in or out of phase depending on how they were reflected from the layers of metal atoms in the crystal. The result was that large numbers of reflected electrons were recorded at some places, alternating with regions at which few or none appeared.

At the same time, British physicist George P. Thomson projected a stream of electrons through a thin film of crystals and observed the alternating bright and dark bands that are the hallmark of wave interference. (As a result, in one of the oddball ironies of quantum science, G. P. Thomson would be awarded the Nobel Prize for showing that the electron was a wave—whereas his father, the famous J. J. Thomson, had received the same honor for demonstrating that it was a particle!)

As experimental evidence accrued, the inconceivable became the incontestable: matter waves really existed. The excitement further deepened the atomic enigma. If electrons somehow behaved like waves, what were they waves *of*? And was there a way to define and predict their behavior?

Matter of Probability

Working independently, the German Werner Heisenberg and the Austrian Erwin Schrödinger set out to answer those questions in the mid-1920s. Both assumed that electrons circling an atom somehow constantly change states, acquiring different values of potential and kinetic energy in each of three dimensions. Those values, of course, are limited in their range by the specific allowed orbits of various types of atoms.

If electrons oscillated in such regular ways, theorists believed, there ought to be methods of expressing their changing properties mathematically, just as there were for describing the way a swinging pendulum alternates between potential and kinetic energy, or the relationship between electric and magnetic fields in electromagnetic waves. Heisenberg and Schrödinger went at the problem from completely different angles using dissimilar methods. Yet each arrived at equations that produced the correct values for specific quantized electron energies as observed in spectral lines—still the benchmark by which any theoretical description would be judged.

Neither, however, could describe in material terms exactly what the electrons were doing. Schrödinger noted that his famous solution, published in 1926, provided the right answers "whatever the waves may mean physically."

What *did* they mean? All hypotheses strained credulity. Schrödinger originally supposed that the electron was somehow smeared out across space around the atom, with most of it in the region where his wave equation generated the largest values. But German physicist Max Born proposed a more comprehensible yet conceptually staggering interpretation: the wave aspect of a particle described the *probability* of its having a certain set of characteristics, such as a specific position, at any particular time. After all, electrons were fundamentally different from pure waves. Although one could observe half a wavelength, or two waves superimposed, one would not expect to see half an electron, or one made of fractions of two electrons. So maybe what the mathematical description of the wave represented was the fractional likelihood of the whole electron being in a given state!

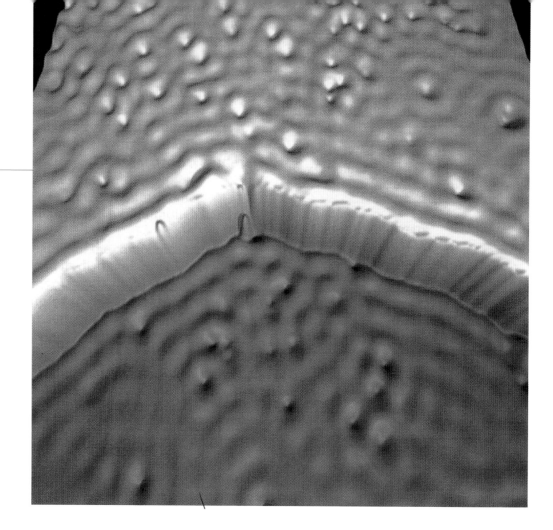

There are many ways to demonstrate the wave character of electrons. Right, a scanning tunneling microscope photograph shows a surface of silver with two step edges. Waves can be seen emanating from the upper and lower edge of the step. These waves are due to electrons that are restricted to the surface of the metal. From the STM electron interference pattern, the probability of various electron motions can be determined and represented visually. Below, the circular pattern indicates that electrons move with almost equal probability in all directions along the surface.

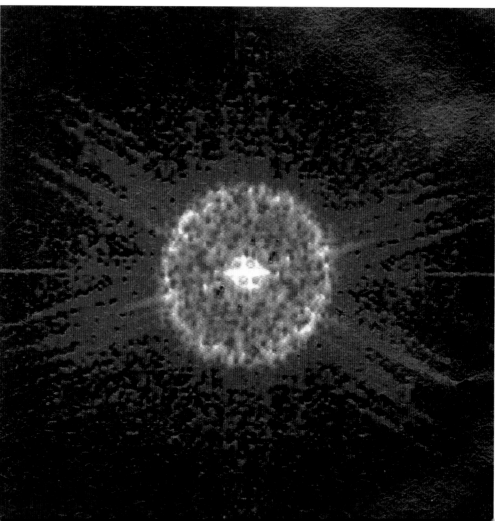

On the silver surface shown on the previous page, electrons are free to move in any direction parallel to the surface. The STM photograph left is an image of a beryllium surface which shows a zigzag step separating two levels. In the beryllium case, the step edges running from top right to bottom left give rise to pronounced wave trains while no waves are induced by the other edges. In this case electrons can move across the surface in some directions much more readily than in others, resulting in a probabilty pattern for the direction of motion, below, very different from the circular pattern, opposite.

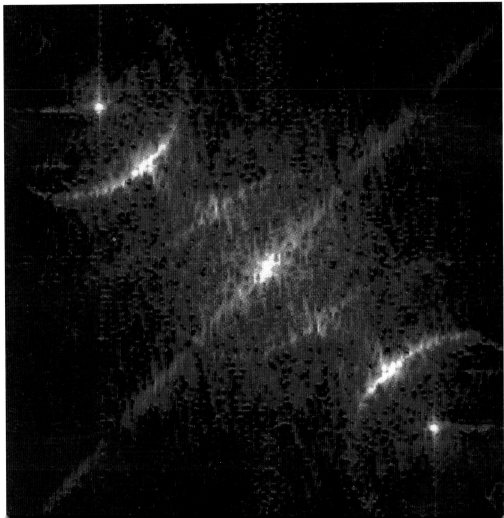

This insight, as expanded and improved by many others over ensuing decades, entailed a stupefying revision of the basic concept of matter on the smallest scale. Contrary to the most fundamental Newtonian precepts, a particle did not have a definite set of properties that, at least theoretically, could be known. Instead, it had a number of possible states that were entirely uncertain until the act of measurement forced the particle to assume one particular, but arbitrary, final condition.

Thus, if exactly the same kind of measurement were made on five electrons with identical quantum characteristics, the observer might get five different values when the range of probabilities "collapsed" into specificity.

At this point, Einstein revolted. "I am at all events convinced," he wrote to Born in 1926, that God "does not play dice." There must, he thought, be some sort of undetected "hidden variables" that determine a particle's properties. Science simply had not found them yet. Seven decades later, none has been seen and none is expected. But Einstein's reaction typified the unease many scientists felt with the growing roster of exotic new quantum notions that did not correspond to any concrete physical phenomena even faintly resembling classical physics. This continues today.

Lines of Inquiry

It hadn't begun that way. Bohr had calculated the allowed energies of the electron in hydrogen atoms by proposing that the electron could only occupy orbits that were specific multiples of a unit of angular momentum. That concept pictured the electron as a tiny electrically charged mass revolving around the nucleus in the same manner as planets orbit the sun. Of course, any moving charge creates a magnetic field. So a negatively charged electron revolving in such a way would be, in effect, a tiny electromagnet and would have its own magnetic field. The combined sum of all of an atom's electron alignments would give it a collective net magnetic field.

That quasi-classical picture worked wonderfully well—up to a point. But there were numerous phenomena that couldn't be completely explained, even for the simplest case of the hydrogen atom. One was an aspect of the Zeeman effect (discovered by Dutch physicist Pieter Zeeman) in which spectral lines are displaced and split when the source atom is placed in a magnetic field.

With Bohr's model, one could make some sense of the Zeeman effect. If an electron is suddenly exposed to a strong external magnetic field, the electron's own intrinsic field will change alignment in response to the outside field. That will alter its angular momentum (and thereby its energy) slightly and in turn change its spectral signature.

However, there were other experimental results that were more difficult to explain. One was the peculiar fact that, when certain spectral lines were examined very closely, they were found to be made up of even numbers of separate, nearly identical lines.

Even more puzzling was the result of a landmark 1922 experiment designed to determine whether—or how much—angular momentum was quantized. German physicists Otto Stern and Walther Gerlach projected a beam of silver atoms through a specially contrived magnetic field of nonuniform intensity. The field strength increased in the vertical direction. As a result, each electron passing through the external field would be deflected up or down by some amount depending on the alignment of the electron's intrinsic field.

Those electrons whose north-south poles were lined up mostly or entirely in the vertical direction would be pushed upward or downward by the largest amount. The net combination of all those forces would nudge the whole atom one way or another. On average, if all the electrons' individual magnetic fields were aligned at random in a continuous range of possible magnetic orientations, the beam should spread into an unbroken vertical line when it hit a detector.

It didn't. Unaccountably, the silver atoms split into two—and only two—separate bunches. This plainly confirmed Bohr's original concept that angular momentum was quantized. But beyond that, it didn't make much sense. If the beam had split into three parts, Stern and Gerlach might have been able to understand this in terms of the quantum theory that existed at that time, but splitting into two parts was totally inexplicable. But within a couple of years, theorists had found an appealing, if startling, explanation: perhaps the electron had yet another kind of rotation. In addition to revolving around the nucleus, it might be spinning on its own axis just as the Earth does while orbiting the sun. If that were the case, some electrons might be spinning clockwise while others were spinning counterclockwise. Each kind of spin would produce a tiny magnetic field aligned in opposite directions, giving the possibility of two, and only two, quantized values. That difference, although small compared to the electron's total angular momentum, might be just enough to account for the multiple, nearly identical lines in atomic spectra.

It might also explain the two clusters of atoms seen in the Stern-Gerlach experiment. If the spin of the silver atom's single outermost electron was quantized, and could be in only one of two states, then the atom beam would split into two sections—precisely what had been observed.

Spinning without Motion

There was, however, an annoying problem with the idea. It was almost certainly impossible in any conventional physical sense. For electrons to spin at the rate required to generate the magnetic quanta observed, they would have to be moving faster than the speed of light. So they weren't really spinning on an axis, even though they clearly behaved *as if* they were; in the same way, physicists were coming to realize, electrons weren't really revolving around the nucleus, even though their apparent angular momentum could be quantified. In those cases, and many more to come, scientists gradually became accustomed to the idea that, in the quantum world, objects had properties for which there were no visualizable physical counterparts.

At the same time, though, increasingly sophisticated mathematical descriptions made increasingly fascinating predictions. In 1928, British theorist Paul Dirac came up with a formulation that accounted perfectly for the spin effects of electrons. (Later in the century, researchers would discover that other elementary particles also possess the perplexing quality of spin.) And it did something else, the profound significance of which was not clear at the time: it permitted solutions for energy values that had a minus sign. Dirac interpreted this to mean that there was some kind of mysterious counterpart of the electron that had "negative energy."

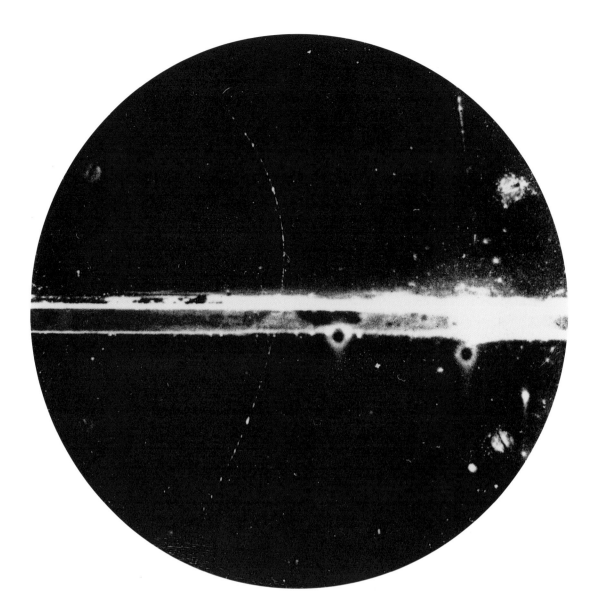

The first evidence for the existence of the positron—the antiparticle of the electron, with the opposite electric charge—was produced by Caltech physicist Carl Anderson in 1932. Anderson had been using a device called a cloud chamber to study the nature of cosmic rays, the fast-moving charged particles that pervade space. In a cloud chamber, low-pressure air contains a quantity of water micro-droplets that are invisible until a charged particle passes through the chamber. The particle imparts an electric charge to the gas through which it moves, causing vapor trails of water droplets to form in the particle's path. The trail shown here is from a positron entering at the bottom and curving counterclockwise. (A clockwise curve would indicate a negatively charged particle.) Note that it bends more sharply after it is slowed by passing through an inch-thick lead sheet that Anderson placed in the chamber.

If a highly energetic photon struck one of these the right way, Dirac reasoned, it would boost it up to a positive condition, leaving a sort of hole where the negative-energy electron had been. Conversely, when a positive-energy electron recombined with a hole, a photon would be emitted.

These completely unfamiliar ideas were, to say the least, a bit hard to take. But in 1932, only four years after Dirac published his equation, American physicist Carl Anderson accidentally provided experimental evidence that Dirac's outlandish idea was right.

Anderson, working at Caltech, was trying to ascertain the nature of "cosmic rays," the high-energy charged particles that rain down on Earth from space. He was using a cloud chamber, an enclosure filled with gas molecules that became ionized when a

charged particle passed through them, leaving a visible trail. The chamber was surrounded by a powerful magnetic field that would cause a charged particle passing through it to curve one way or another. The direction of the trajectory would reveal the charge; the degree of curvature would indicate the mass.

Anderson started taking data, and swiftly made a startling find: The particles were both negatively and positively charged. The negative objects were clearly electrons. But what were the positive particles?

The curvature showed that each had the same mass as the electron, but an unaccountably opposite charge. Anderson had discovered the anti-electron (later named the positron)—the first antimatter ever detected.

Cloud chamber photographs can determine the approximate mass and electric charge of a particle. In general, the slower a particle moves through the chamber, the larger the vapor trail it makes. The charge is detected by the way the particle's trajectory bends as it passes through a magnetic field: particles of opposite charges will be deflected in opposite directions. Carl Anderson is shown with the coil used to create the magnetic field pervading the cloud chamber.

Physics by the Numbers

Meanwhile, other theoretical advances were arriving with dizzying speed. Although it was often impossible to generate a mental picture of particle properties, physicists had learned to characterize them in appropriate ways using "quantum numbers," a set of values that described each electron's orbital "shell" or energy state, its orbital angular momentum, the orientation of that momentum in a magnetic field, and finally its spin.

Using these concepts, Austrian physicist Wolfgang Pauli made a momentous claim in 1927: no two matter particles in an atom can have identical quantum numbers. This "exclusion principle," simple as it sounds, would turn out to have spectacular explanatory power.

For example, it would make comprehensible the magnificent, but then inscrutable, regularity of the periodic table of elements devised in the nineteenth century by Russian chemist Dimitri Mendeleev and still in use today. Mendeleev and others had observed that, when all the elements were arranged in a list by weight, those with similar properties (such as sodium and potassium, or chlorine and bromine) seemed to recur at predictable intervals, or periods.

But why was the table periodic? And what determined the length of the periods? Pauli's exclusion principle explained both. As atoms got larger, they filled each successive energy level or "shell" of electrons until adding another would put two electrons in the same quantum condition. At that point, the electron had to go into the next shell. The number of electrons in the outermost unfilled shell determined the element's reactive properties. Chemistry had become a quantum affair.

One or the Other and Both

But for every puzzle that quantum mechanics solved, it seemed to generate a dozen more. And nothing caused more cerebral consternation than the stubborn indeterminacy of particle behavior, typified by Heisenberg's uncertainty principle. Heisenberg had demonstrated mathematically that, for certain pairs of characteristics such as momentum at a specific position, or energy content at a precise time, the value of one variable could only be known at the expense of a corresponding uncertainty in the other.

Another way to look at the issue is that it is impossible to measure something without disturbing it a little. Putting a ruler next to an object nudges it slightly; shining a light on an object bombards it with photons that have momentum. In the macroscopic world, those effects are negligibly small. But at the quantum scale, they are sufficient to alter the state of the particle dramatically.

For example, in order to measure the position of an electron accurately, you have to bounce a photon off of it. And to get a precise measurement, you need a photon with a very short wavelength. But as Planck and Einstein showed, the shorter the wavelength, the greater the energy that a photon carries. So although the measuring photon might provide an exact value for the instantaneous position of the electron, it would transfer considerable energy to the electron, thus changing its momentum at the same time.

The uncertainty principle in particular, and the fundamental indeterminacy of quantum behavior in general, might seem like appalling liabilities for experimentalists. But in fact they can be employed to create novel and classically impossible states of matter. In one case, cooling an atom close to absolute zero reduces its thermal motion, thereby greatly reducing the uncertainty in its momentum. But as its momentum becomes more certain, the wave function that corresponds to its position gets larger—large enough, in fact, for the waves of different atoms to overlap. They form a sort of quantum "superatom" called a Bose-Einstein condensate, which gets part of its name from Indian physicist Satyendra Nath Bose, who independently developed much of the theory needed to predict this effect.

Early indications of this condition, predicted in the late 1920s, were seen eventually in the low temperature superfluid state of the element helium and in certain other experiments. But Bose-Einstein condensation in its purest form was not achieved in the laboratory until 1995 because of the formidable technical obstacles to slowing and "trapping" atoms. Eventually, scientists at the U.S. National Institute of Standards, Stanford University, and the École Normal Supérieure in France succeeded by using a combination of magnetic fields and "laser cooling" (in which an incident photon is used to slow the motion of an atom) to snare a bunch of atoms at less than one millionth of a degree above absolute zero, the lowest temperature theoretically possible. In that condition, the wave functions of the atoms begin to overlap, producing Bose-Einstein condensates, peculiar "atom laser" effects, and other oddities.

Yet another aspect of quantum indeterminacy, called "tunneling," has formed the basis for several modern practical applications. Because the nature of any particle is innately probabilistic, there is a small but finite possibility that it can penetrate and traverse a barrier through which it would not normally be allowed to go.

Russian-born physicist George Gamow had used that idea to explain how alpha particles emerge from the nuclei of radioactive elements. In 1962 British physicist Brian Josephson predicted that tunneling should cause remarkable effects at a thin insulating barrier between two superconductors—that is, materials that have been cooled to such a low temperature that they lose all resistance to current flow. Josephson determined that pairs of electrons would tunnel across the junction between two superconductors even when there was no voltage difference between them. When there was, the tunneling pairs would oscillate back and forth, with a frequency dependent on the voltage. The presence of a magnetic field (which would induce a small current in the superconducting material) would therefore cause the system to fluctuate in a predictable way, making the Josephson junction an exquisitely sensitive detector of fields.

A further application arose in the early 1980s, when Gerd Binnig and Heinrich Rohrer invented the scanning tunneling microscope described in Chapter 1.

Another particularly eerie aspect of quantum mechanics results from Heisenberg's omnipresent uncertainty principle: the more certain the time interval in a quantum

system, the less certain the energy content, and vice versa. So if the time increment is extremely brief—on the order of trillionths of trillionths of a second—and hence enormously certain, the energy content becomes correspondingly uncertain.

That energy can manifest itself as "virtual" particle-antiparticle pairs. Consequently, the ostensibly barren vacuum actually is seething with pairs that wink in and out of existence on the narrowest time scales. "Empty" space is by no means empty.

Such phenomena might seem so nearly mystical as to be easily disregarded in practice. But they are not. The effect of such vacuum fluctuations on electrons surrounding hydrogen atoms is large enough to cause a discernible shift in spectral lines, as American physicist Willis Lamb discovered in 1947.

And in fact, the existence and function of virtual particles form an integral part of quantum electrodynamics (QED), the theory of how electromagnetic forces and matter interact by exchanging virtual photons. QED, developed independently in the 1940s by Americans Richard Feynman and Julian Schwinger, and Japanese physicist Sin-itiro Tomonaga, produced calculated values that were a perfect fit with the experimentally observed Lamb shift.

Spooky Action

Nearly a century after its advent, quantum mechanics is still evolving. Theorists are still struggling to extend its concepts to additional phenomena, notably gravity. Experimentalists are still looking for the waves associated with gravity, while turning up evidence of other peculiarities. And both are still trying to understand its ramifications.

Einstein never reconciled himself to the fundamental indeterminacy of quantum mechanics, which despite its successes he regarded as a woefully incomplete explanation of nature. In one famous analysis published in 1935, he and theorists Boris Podolsky and Nathan Rosen attempted to show that the theory entailed a fatal internal contradiction. Suppose, they argued, that a process creates an identical pair of particles which fly off in opposite directions until they are millions of miles apart.

Quantum theory insists that neither particle has a definite position or momentum until it is measured. So if someone finally makes a measurement of the momentum of Particle 1, it simultaneously determines the exact momentum of Particle 2.

But because the two particles are now so far away from each other, no interaction or information transfer could have occurred between them, even at light speed. That is, nothing could have "told" Particle 2 what momentum value to have prior to the measurement. Thus by any traditional definition of physical reality, Particle 2 should have had that value *before* the measurement. Since quantum mechanics forbids that outcome, the three argued, the theory must be inadequate.

The intellectual battle lines were drawn between Einstein partisans, who didn't like the idea of two distant particles correlating with each other instantaneously, and the supporters of the "Copenhagen interpretation," which takes its name from Niels Bohr, who, more than anyone else (even Schrödinger and Heisenberg), organized the intel-

American theorist Richard Feynman had an enormous impact on mid-century science, making basic contributions to many different fields of physics. He is best known as one of the three principal authors of quantum electrodynamics, a comprehensive theory of the interaction between electrons and photons. But he was also a captivating personality who, through numerous public events, television appearances, and popular books, served as an envoy from the often bewildering world of modern physics to the culture at large.

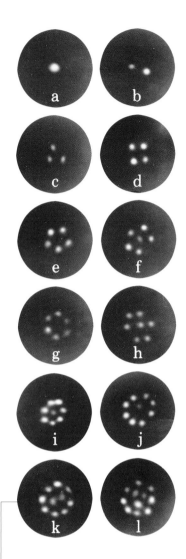

Quantum physics, which describes how energy is conveyed in discrete units called quanta, has led to many bizarre predictions and observations—most of them in the domain of extremely small dimensions. Yet on rare occasions, physicists have seen evidence of quantization in relatively large-scale systems. These images, made in 1948 by Richard Packard of the University of California at Berkeley, show quantized whirlpools or vortices in a rotating container of liquid helium. The helium has been made so cold that it becomes a "superfluid"—that is, it loses all resistance to flow. Packard devised a method of making the quantized vortex phenomenon visible, and the resulting images show how the vortex pattern changes when the superfluid helium container rotates at different speeds.

lectual framework in which quantum mechanics would be used up to the present day. In its broadest terms, the Copenhagen position insists that it is pointless and misleading to try to reconcile quantum observations with conventional notions of reality. Quantum entities do what they do; that's all we can say for sure.

For years, some of the best minds in physics and math pondered this problem, including American David Bohm and Irishman John Bell. Hungarian-born mathematician John von Neumann weighed in, as did numerous experimentalists anxious to confirm or disprove Einstein's premise. Finally by 1982, laser technology had advanced to the point at which French physicist Alain Aspect and colleagues at the University of Paris—using pairs of photons instead of particles—showed that the orthodox quantum interpretation seemed to be correct: each photon in a pair *did* have exactly complementary characteristics even when measured in a time interval too short for light to have traveled between them. Quantum observations simply do not agree with conventional reality: two distant (entangled) particles *can* apparently communicate with each other instantaneously, Einstein's views to the contrary notwithstanding.

Of course, the outcome by no means resolved the issue conclusively, and physicists will continue to explore that and other quantum puzzlements into the twenty-first century. Thus in many respects, an observation of Feynman's is still true. "I think I can safely say," he once remarked, "that nobody understands quantum mechanics."

Yet it remains the most accurate scientific theory ever constructed to explain the physical world, with predictions confirmed by experiment to 10 significant digits. And its myriad ramifications continue to suggest new lines of research.

For example, physicists recently determined that quantum uncertainty itself may be exploited in ways that could lead to dramatic advances in computing. Conventional computer processors are sequential devices that can only execute one instruction at a time. As a result, certain kinds of problems—such as trying out possible factors for large prime numbers—are enormously time-consuming.

But the inherent indeterminacy of quantum systems, such as an electron in an atom, means that at any given instant a particle is in a combination or "superposition" of many possible states at the same time. That condition, physicists began to suspect in the 1980s, might be used as a way to compute a number of possibilities simultaneously. Such "quantum computers," theorists believe, will eventually be able to do certain kinds of calculations in a fraction of a second that would take traditional digital machines weeks or months to complete. Building that kind of device is expected to be one of the most exciting challenges for physicists in the coming century.

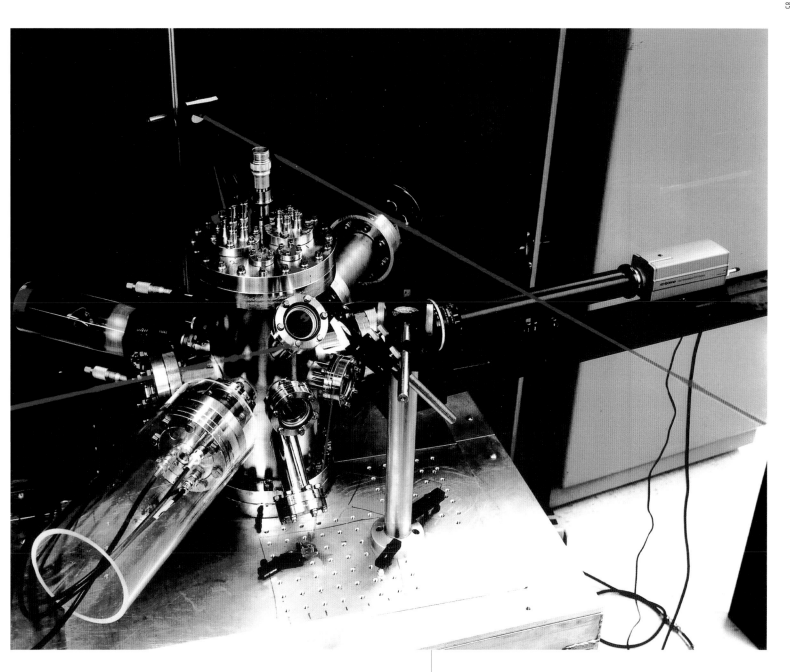

Conventional electronic computers process information that has been converted into binary digital form. That is, each unit of information is represented by strings of 1s or 0s. Those two states are embodied in the computer within tiny circuit elements that are either on (1) or off (0). But in a quantum computer—which physicists hope to build some day—an element could be in both conditions at once! That's because quantum mechanics recognizes that objects can exist in a "superposition" of two different states simultaneously. Thus, certain kinds of calculations could be performed much faster. Scientists exploring that possibility have built devices such as this one, which uses laser light to switch calcium ions between two states.

These scanning tunneling microscope images, from IBM's Almaden research center in California, show quantum behavior in stunning visual detail. The image above shows the surface of a copper crystal on which 48 atoms have been placed and individually positioned to form a circle. Like most metals, copper crystals have a structure that allows some electrons to flow freely among the outermost atoms. These "surface state" electrons cannot penetrate into the crystal, but travel along the surface like a sheet of water flowing down a window. When the electrons encounter an obstacle, such as the circle of iron atoms, they are partially reflected backwards. Interference between the waves trapped within the iron atoms form circular ripples while those on the outside produce dimpling.

The STM image on the opposite page shows a "landscape" on a bit of copper a few thousand atoms wide, with a crystalline topography revealing various layers of a geometrically regular arrangement. As quantum mechanics dictates, the surface electrons behave like waves, creating the eerie ripple effect seen in this image. Each individual ripple is about 10 atomic diameters wide.

4 STRUCTURE

By the late 1920s, quantum theory—despite its unsettling ramifications—was firmly established. Yet it was a triumph tinged with dismay. The inherent indeterminacy of quantum states, and the fact that they took on definite values only in the presence of an observer, seemed to threaten the end of human comprehension. Heisenberg himself intoned that "the conventional division of the world into subject and object, into inner and outer world, into body and soul, is no longer applicable." Harvard mathematician and philosopher P. W. Bridgman went even further. Quantum principles meant "nothing more nor less than that the law of cause and effect must be given up," he wrote in 1929. "The world is not a world of reason, understandable by the intellect of man."

As it turned out, that sentiment was spectacularly wrong. It was the beginning of an age of unprecedented understanding, in which physicists would extend the precepts of quantum mechanics beyond single atoms to describe and control the structure and behavior of large-scale materials. Quantum theory, they found, could account for many aspects of substances, including color, hardness, crystalline form, optical behavior, and magnetic characteristics.

But perhaps most important to everyday life in the twentieth century, it could explain the microscopic electrical properties of materials, a critical part of the new field of "solid-state" (that is, solids) or "condensed-matter" (solid and liquids) physics. Physicists exploited those properties at an accelerating pace that has yet to slow.

Fewer than fifty years after the quantum revolution, modern civilization was running on quantum-mechanical devices called transistors. And by the end of the century, science would be well on its way to describing astonishing phenomena inconceivable in the physics of Faraday or Maxwell, including resistance-free superconductors and laser light.

One of the major accomplishments of materials scientists has been the creation of non-metallic substances that have many of the characteristics of metals but are much lighter and can tolerate much higher temperatures. This scanning electron microscope picture shows a thick fiber of silicon carbide that has been heated to 1,300 C and exposed to nitrogen gas. The process caused the fiber to grow reinforcing "whiskers" of silicon nitride. The experiment was one of a series attempting to improve the toughness and strength of ceramic composite materials.

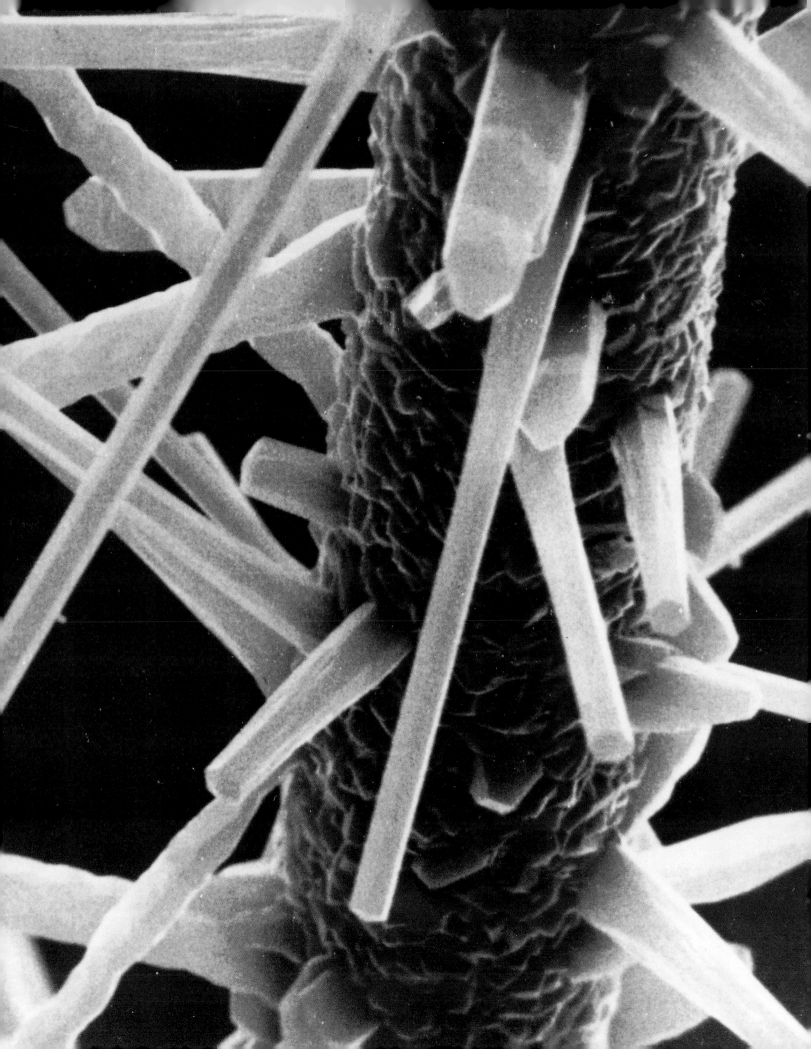

The precursor to the transistor was the vacuum tube, described in Chapter 2. In 1906, American researcher Lee De Forest applied for a patent on a "device for amplifying feeble electric currents." That is a succinct description of what radio receivers do. A radio signal induces a faint pattern of alternating currents in the receiver's antenna. Detecting that pattern is fairly easy; but boosting the tiny currents to a level at which they can power headphones or speakers posed a serious problem. De Forest's solution, right, was called a "triode" because it contained three electrodes. The weak radio signal currents were run through a control electrode placed between two high-voltage electrodes, all of which was surrounded by a glass vacuum tube. In the absence of a signal, a current could almost—but not quite—pass across the vacuum gap between the high-voltage anode and cathode. But when charges moved in the control electrode, they provided just enough attraction to permit a current to flow between the two high-voltage electrodes. Thus the pattern of the feeble signal from the radio antenna was imprinted on a current strong enough to drive audio devices.

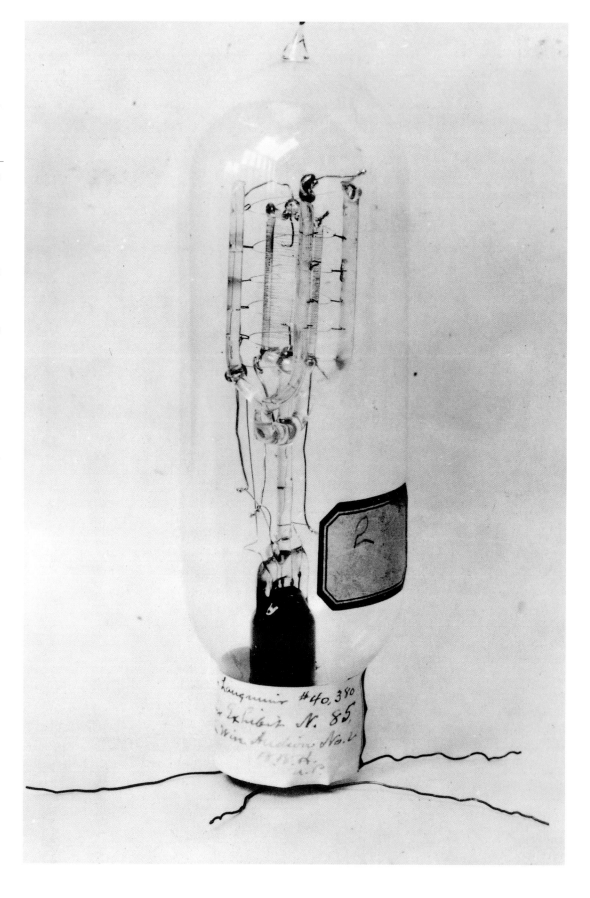

Conduction in Solids

The discovery of the electron was a momentous revelation. But it didn't answer the key questions involved in making practical electrical devices. How, exactly, are electrons conducted in metals? And what makes a good conductor? The most promising explanation was propounded by German physicist Paul Drude and others around the turn of the twentieth century: many of the electrons in a metallic solid behave rather like the particles of a free gas. Each is somehow detached from its parent atom, and each has a different, random velocity—typically in the range of hundreds of miles per hour.

In the absence of a voltage, these free electrons just ricochet around, banging into each other or into the geometric array of atoms that make up the metal lattice. But when a voltage is applied, each free electron gets a non-random shove in the direction determined by the electric field the voltage creates. Thus the whole zigzagging multitude of electrons drifts the same way, slowed by frequent atom collisions, at a collective speed on the order of inches per second. That was the gas model of current "flow."

The gas model was strikingly successful in explaining many observed properties of metals, but it failed miserably in others. Notably, it could not account for the well-established dependence of electrical conductivity on temperature, nor for the puzzling fact that metals lose resistance when cold, whereas substances such as silicon conduct more current the warmer they get!

The vacuum tube rapidly became the world's premier electronic device during the first half of the century. Naturally, it became the foundation for early electronic computers such as the "differential analyzer" shown here. Built at Massachusetts Institute of Technology around 1945, it filled the better part of a room. But it had less computing power than many hand-held calculators available today.

In addition, a gas of free electrons should be capable of soaking up a lot of thermal energy, with the result that metals should be able to absorb a great deal of heat per degree of increased temperature. But careful experiments showed that was not the case.

With the advent of quantum theory and the exclusion principle, those problems began to yield. Physicists realized that electrons in a given material, just as in any single atom, could not exist in an infinite range of energy states, as the free-gas model implied, but only in a number of specific quantized energy conditions. And when they gained or lost energy, they could not occupy a continuous spectrum of positions; instead, they could fill only those positions not forbidden by the exclusion principle.

Quantum theory also explained why 100 silver atoms bound into a solid crystal behave quite differently from 100 separate silver atoms. In bulk materials, atoms are so close together that their electron orbitals overlap and merge as electrons are shared. These overlaps create "bands" of possible energy values that an electron can have.

Electrons nearest the nucleus are tightly bound to their positions. But electrons in the outermost bands are less constrained and can more easily occupy a number of energy states. In many substances, these outer electrons can migrate from place to place within the solid. And just as each kind of individual atom has particular orbits that are not allowed (the effect that causes the spaces between spectral lines), bulk materials have whole ranges of electron energies that are forbidden by quantum rules and the exclusion principle. These energy ranges are called "gaps."

Gaps were first described by the Dutch physicist M. J. O. Strutt in 1928, and shortly thereafter Felix Bloch and colleague Rudolf Peierls used similar ideas to describe the behavior of electrons in metals. By 1931, British theorist Alan Wilson had applied the band-gap concept to explain how metals differed from insulators and from the peculiar class of partial insulators called semiconductors.

Metal solids, according to the new theory, have a sizable upper band of permitted electron energies (now called the "valence" band) that is not even close to being filled with electrons. So when something, say an electric field or thermal motion, adds energy to electrons in the metal crystal, there are empty states that they can occupy without violating the exclusion principle which prohibits any two electrons from being in exactly the same quantum condition. If a field is present—as when the two terminals of a battery are attached to a metal wire—a majority of those electrons can move in the direction dictated by the field because unoccupied spaces are available. That is, a current can flow. Indeed, quantum conditions in metals allow the valence electrons to travel freely around the crystal, unattached to any particular atom, like sea water flowing between islands.

Materials with just enough electrons to fill the valence band cannot conduct electricity, because they contain no unoccupied states and there are the same number of electrons moving in one direction as in the opposite direction. These materials, called insulators, also have no readily accessible place for electrons to go beyond the valence band: the gap between the valence band and the next permitted band is huge. Thus, under ordinary circumstances, the electrons in a piece of wood or chunk of diamond do not have sufficient energy to rise into the next available band, which would be necessary in order for them to move around. As a result, no current flows.

Semiconductors, Wilson showed, also have full valence bands. But they have a permitted energy band, now called the conduction band, that is only a short gap away. So only a modest amount of increased thermal energy causes some electrons to jump up to the conduction band, enabling current flow. Among other things, the band theory conveniently explains why semiconductors allow more current flow the warmer they get: the more thermal energy the material receives, the more likely it becomes that an electron will have enough energy to get into the conduction band, enabling a small current to flow.

Bridging the Gap

Wilson also arrived at a crucial insight that would power the transistor explosion fifteen years later: "The observed conductivity of semiconductors," he declared, "must be due to the presence of impurities." This deceptively simple notion—refined and extended during the 1930s and 1940s, and still of intense interest today—lies at the heart of every modern radio, TV set, and computer, as well as hundreds of kinds of miniaturized electronic devices.

The impurity theory made good quantum sense. If foreign atoms are inserted into an otherwise pure crystal of some semiconductor, such as silicon or germanium, they alter the behavior of the electrons, and thus inevitably change the collective band structure of the material.

A silicon atom, for example, has four electrons in its outermost shell. When billions of them combine, they form a geometric crystal lattice in which each atom shares an electron with four neighbors. But an element from the next column to the right in the periodic table, such as phosphorus, has five electrons in its outside layer. Thus if a phosphorus atom is placed in the silicon lattice (a process called "doping"), there is one extra electron with no particular place to go. Its energy state is higher than the silicon valence band, though lower than the conduction band. So less thermal energy is needed to boost electrons into the conduction band. The more phosphorus present in the crystal, the more current it will conduct when a voltage is applied. These materials were called "N-type" for their majority negative charge.

Doping can also produce the opposite effect if the impurities come from elements one column to the left of silicon in the periodic table. Those elements, such as boron, have only three electrons in their uppermost levels. So when a boron atom is introduced to a silicon lattice, it shares electrons with three adjacent silicon atoms, leaving a space or "hole" where the fourth electron ought to be.

This hole, because it represents the absence of a negatively charged electron, acts like a unit of positive charge. It can migrate around in the crystal just like a real particle in response to electric fields, and performs as a charge carrier. Such positively doped substances were called "P-type."

As this new understanding was emerging, so were a number of practical problems in the fast-moving telecommunications field. Engineers desperately needed improvements in several kinds of devices. One was the rectifier used to convert the AC signal of radio waves into DC current that could drive loudspeakers and headphones. Another was a dependable amplifier or "repeater" to boost the strength of signals when they faded after traveling over hundreds of miles of wire.

As we saw in Chapter 2, the electronics industry had come to depend on "vacuum tube" diodes and triodes for rectifiers and repeaters. Inside these clever devices, the current stream originated in an electrode made so hot that it emitted electrons. They were not terribly dependable. The slightest defect in the glass enclosure would spoil the vacuum, and the intense heat generated by the electrodes caused the tubes to burn out at unacceptably high rates. By the 1940s, many researchers, including a newly assembled team of physicists at Bell Telephone's research laboratory outside New York City, had begun an intense search for more durable substances and processes.

Controlling Resistance

Early attention had focused on the junction between layers of different materials—copper and copper oxide, for example. Experiments had shown that such junctions could function as rectifiers because they only allowed current flow in one direction (from the metal into the oxide) and not in the other. As a result, they could convert an alternating current signal into a direct current signal without the need for bulky tubes. The problem was that nobody knew exactly why that happened.

Several researchers, notably including Bell's Russell Ohl, had been trying to understand and control what happened at the junction between two such materials, but World War II delayed most of the work. In 1945, however, Bell Labs launched a major experimental effort, convinced that the future of telecommunication electronics lay in solid-state devices. One team, containing William Shockley, John Bardeen, and Walter Brattain, set out to investigate the problem.

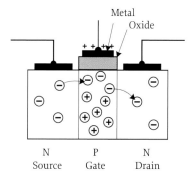

Metal
Oxide

N
Source

P
Gate

N
Drain

They tested various combinations of P-type and N-type junctions under numerous conditions, hoping to find, among other things, a configuration that would allow a thin layer of one or another semiconductor type to regulate a large current flow between two electrodes. In effect, they were searching for the electronic equivalent of the way a canal lock works: a relatively low-power, easily manipulated device (the movable lock gate) controls the transfer of huge quantities and pressures of water from one part of the canal to another.

What the physicists needed was a reliable process whereby a very weak current pattern—such as an attenuated telephone or radio signal—could be used to vary the resistance of the middle semiconductor layer, just as a small amount of energy can raise and lower the canal-lock gate.

In theory, researchers knew, that should be possible without any mechanical devices by exploiting the properties of doped semiconductors. Suppose, for example, that one wanted to control the current flow between two pieces of N-type semiconductor, each of which was connected to an electrode in a circuit. (The N-type piece with a surplus of incoming electrons is called the source; the piece on the other side is called the drain.) If the source and drain were placed in direct contact, current would flow easily through both pieces, completing the circuit.

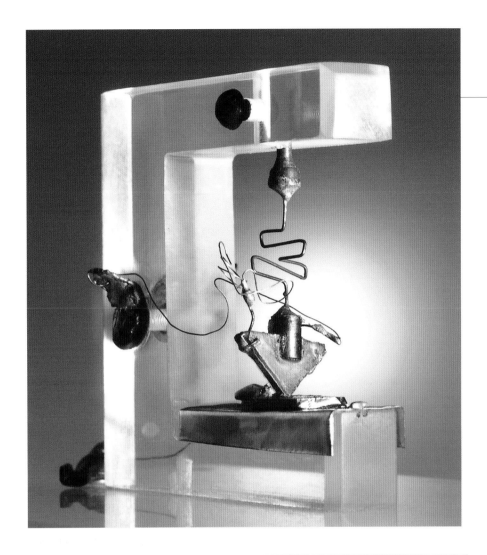

Although individual transistors are now rarely larger than a few millionths of a meter, the original model was a couple of inches wide. Shown here is the famous experimental device built by John Bardeen, William Shockley, and Walter Brattain, and demonstrated on Christmas Eve of 1947.

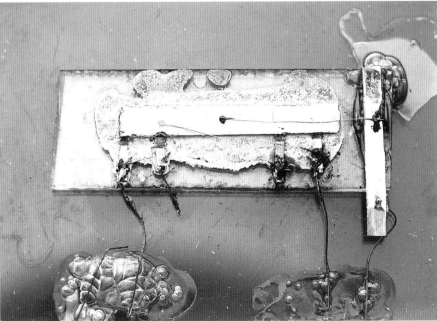

In the earliest transistorized electronic circuits, individual components were connected by wires. Then in 1958, Jack S. Kilby of Texas Instruments devised the first integrated circuit in which all the components of a device were embedded and interconnected on the same silicon plate. That invention, called the integrated circuit—the first version of which is shown here—made further miniaturization of electronic elements possible.

The invention of the transistor in 1947 revolutionized electronics. And the subsequent ability to combine many transistors into "integrated circuits" eventually made it possible to create extremely intricate logic circuits in a very small space. This 1966 photo shows a development prototype of the first Texas Instruments hand-held calculator. All of the wired components shown here were necessary for the encircled calculator (bottom right) to work. Shortly thereafter, the electric circuits could be contained within the calculator itself.

If the two N-type pieces were separated by a layer of P-type semiconductor (called a gate), exactly the opposite would happen: no current would flow across the N-P junction. That's because electrons traveling from the source into P-type gate would be caught in the holes of the gate material and would not continue to travel into the drain.

But if a tiny electrode was placed on the gate, it could function as a switch. A small positive charge on that electrode would repel the positively charged holes in the P-type semiconductor gate, which would tend to gather in a part of the gate material farthest from the electrode. With the holes out of the way, the current would then be free to pass between the source and the drain. But when the charge on the gate electrode was switched off, the holes would return to their normal positions, preventing current flow.

Thus even a small charge at the gate could control the movement of a large current between the source and drain. Whatever pattern of charges was applied to the gate would be impressed on the much stronger current flow of the source-drain circuit. That is, the gate signal would be amplified.

Finally, in December 1947, the Bell team succeeded in making an ungainly but marvelous device that demonstrated this junction process across 0.002 inch of P-type germanium. Soon they were confident enough to announce a demonstration. As officials looked on, the anxious researchers spoke into a microphone circuit routed through the germanium contact—and heard amplified voices on the other end of the line! Measurements showed that the output was about 18 times as strong as the input signal. They had fashioned a functioning transfer-resistor, or transistor, the world's first solid-state amplifier. Electronics would never be the same.

The team went on to create a practical variation dubbed a "bipolar" transistor. Almost immediately, Bardeen conceived of a somewhat different design called the "field-effect" transistor (the type most widely used today) that had been suggested by earlier work of Brattain. Both kinds of transistors could also be used as miniature high-speed electrical switches and thus were ideally suited to fill the emerging need for digital information processing.

The first version of the modern electronic computer was built in the basement of the physics department at Iowa State University in 1939 by theoretical physicist John Atanasoff. He built the second version, shown here, in 1941. A quarter of a century later, various parties claimed credit for being the first to construct a working computer. A lawsuit ensued, and eventually the court ruled in favor of Atanasoff. No patent rights were granted, however, because of the amount of time that had passed.

In 1985, Intel introduced the first processor chip to contain 1 million transistors, drastically increasing the power of desktop personal computers. Although the complete unit, with all its plug-in prongs and various connectors, was about the size of a silver dollar, miniaturization techniques made it possible to place all the processor circuits (shown in this photo) on a silicon sheet no larger than a fingernail.

The Binary Boom

Digital or binary systems represent numbers and letters as strings of binary digits (0s and 1s) that correspond conveniently with two entirely unambiguous electrical states: on or off, charged or uncharged. But how, computer engineers wondered, could they store these "digital" data in a rapidly retrievable form? Happily, nature had already provided the answer in the form of magnetism, another phenomenon with a fundamentally binary form. All magnets are dipoles, with opposite north and south poles. Thus their orientation (north up or south up, say) could be used to record units of digital information.

That would not have been possible had it not been for a half century of intense research on the microscopic nature of magnetism. Physicists had long observed that some materials were more easily magnetized than others (that is, were more "ferromagnetic"), and had found that their magnetic capacity declined as they got hotter. In 1895 French researcher Pierre Curie showed that each element loses all its magnetism above a certain point now called the Curie temperature. But it was not until quantum theory demonstrated that each charged particle had a specific magnetic field associated with it that physics was able to explain magnetism on the atomic scale.

Recall that the Pauli exclusion principle usually results in pairs of electrons whose spins are opposite; each cancels out the other's magnetic field. But in many kinds of atoms, physicists determined, the electron orbitals arrange themselves so that spins do not cancel out. Iron, for example, has four unpaired electrons in one subshell, giving each iron atom a strong net magnetic property. In large arrays of atoms, these net fields tend to line up in the same orientation. Like any other physical system, they naturally assume a configuration that requires the least amount of energy to maintain; for neighboring clusters of atoms, that means having their fields parallel.

A crucial further insight came from French physicist Pierre Weiss, who determined in the 1920s that clusters of atoms with the same net local magnetic orientation tended to arrange themselves into "domains" a millimeter or so in diameter, with each domain generally aligned at random. How easily a substance could be magnetized, researchers found, depended on how difficult it was for the walls that separated domains to move as their constituent atoms flipped around to line themselves up with an external field. The domain concept also explained the Curie temperature relation: At some point, motion from thermal energy would become large enough to keep the domains randomized.

Once those key concepts were understood, physicists were able to discriminate with unprecedented mastery between the two kinds of magnetic materials needed for a dependable recording apparatus: "hard" materials that were prone to hold their magnetic configuration and thus were ideal for use in recording media such as tape or computer disks; and "soft" materials whose magnetic orientation could be easily and rapidly reversed, making them suitable for recording heads, which had to imprint a fast-changing pattern of data on the storage medium. In 1951, magnetic tape made its computer debut in UNIVAC, the first digital computer produced for sale. Within a few decades, it was the preeminent memory system for audio and video.

In the mid-1990s, Intel announced the Pentium chip containing 5 million transistors in an area about the size of a dime. Techniques of etching fine lines into silicon had progressed to the point at which individual circuit lines, or traces, were only a fraction of a millionth of a meter wide.

The conventional transistor is turned on and off, like a switch, by electrons. Today's computers have millions of transistors in them and each transistor requires about 1,000 electrons to turn it from on to off. This photograph shows gold electrodes on a surface of gallium arsenide forming a single-electron transistor. As the name implies, the single-electron transistor can be switched from on to off by a single electron.

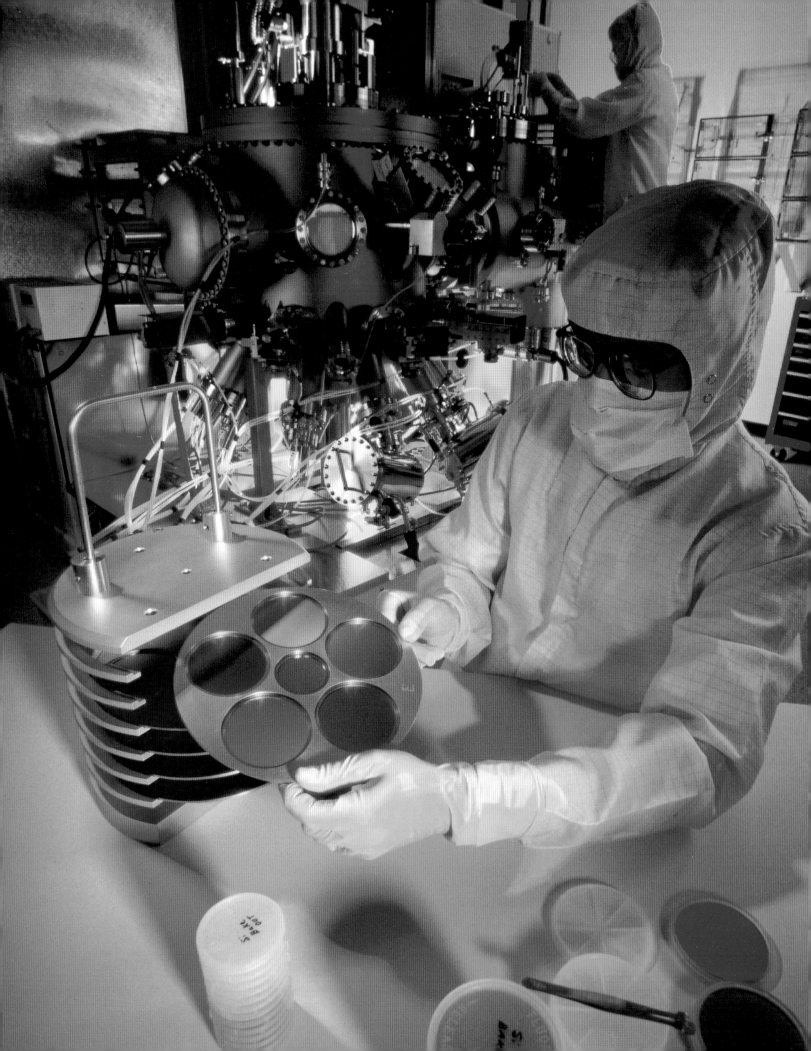

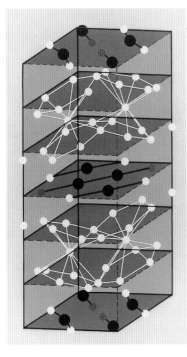

In the search for new materials, strength and lightness are often priorities. The alloy called Magnequench, developed in 1984, is capable of sustaining a strong magnetic field and is used in making lightweight magnets. It is made using a "melt-spin" technique (above). The schematic diagram left illustrates the structure of the crystal. Each unit cell contains fifty-six iron atoms (yellow), eight atoms of neodymium (black), and five atoms of boron (red). The presence of boron, the fifth-lightest element, helps reduce the weight of the alloy.

Although silicon remains the semiconductor of choice in most chip-making applications (because it is abundant and therefore cheap, and is also relatively easy to grow in large crystals), several other elements have proven extremely valuable. One is gallium, a cousin of aluminum. Compounded with arsenic in the form of gallium arsenide, it is able to switch states much faster than conventional silicon structures. Thus, it is much in demand for telecommunications devices that process high-frequency signals. The equipment shown here is used to produce high-quality gallium arsenide circuits.

Magnets behave as if they contained only two separate poles, called north and south. But physicists have discovered that ferromagnetic materials—that is, those that can be made into magnets, like iron—are usually made up of many individual magnetic areas, called domains, in which tiny clusters of atoms have a common magnetic orientation. So when an object is magnetized, a majority of its domains become aligned with an external magnetic field and can remain aligned for a long time. Otherwise, the domains may have a random alignment, such as the pattern shown here on the surface of a cobalt crystal. The image represents an area about 18 millionths of a meter wide.

This scanning electron micro-scope image reveals the magnetic domain structure of an amorphous magnetic ribbon, with different colors representing different mag-netic orientations. Such rib-bons are used at the cores of power transformers. The interesting pattern is caused by a defect in the center of the image. Defects can lead to energy losses.

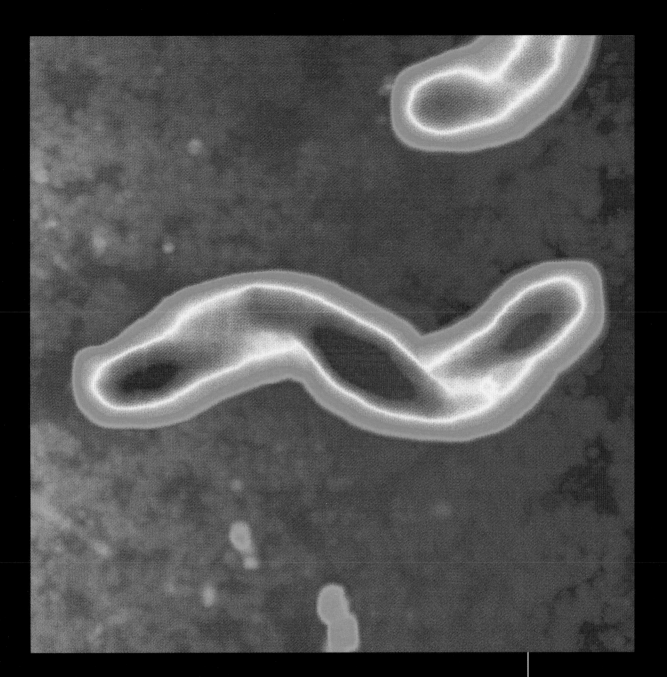

Some bacteria, called magnetotactic bacteria, produce chains of permanent magnetic particles internally. This allows them to orient themselves in the Earth's magnetic field, using it to guide them as they migrate to favorable habitats. This is a magnetic force microscope image of such a bacterium.

Here a scanning electron microscope shows magnetic domains on the surface of an iron-silicon crystal with 3 percent silicon impurities. The four colors correspond to the four prevalent orientations of the magnetic domains. The area shown is about 100 millionths of a meter wide.

The Geometry of Solids

A scanning tunneling microscope image reveals the crystalline surface of silicon. When the surface is miscut by a few degrees, the surface reconstructs forming rows of silicon atom pairs, called dimers, on terraces separated by one monolayer. Three terraces are shown with dimers running perpendicular to the steps. The mean terrace dimension is approximately 40 angstroms.

As understanding grew, physicists were able to classify the principles of atomic organization in extraordinary detail and learn how the structure of composite materials determines their characteristics. One familiar example is magnetoresistance—the way certain materials change electrical conductivity when exposed to a magnetic field.

Controlling that property made possible the "heads" that float over hard disks. These super-thin materials "read" magnetically embedded data by constantly changing resistance, thereby converting the disk's pattern of tiny magnetic fields into a pattern of current changes.

Other researchers found ways to create absolutely regular crystals, and thus avoid the risk of stress fractures and other problems caused by faults or dislocations in crystals. Soon they were turning out perfect silicon crystals for computer chip making, and "growing" high-speed turbine blades out of a single, perfectly homogenous alloy crystal.

Along the way, the new comprehension of natural and artificial crystal structures revealed patterns that were both physically and mathematically beautiful. Entirely unexpected atomic bond geometries were discovered, including a category of "quasicrystals" with patterns that were regular, yet never exactly repeated, many of which were subsequently explained by British theorist Roger Penrose.

An examination on the atomic scale shows that atoms on surfaces often change their positions relative to what would be expected from the underlying crystalline structure. This effect is called "surface reconstruction." This STM image shows how individual silicon atoms arrange themselves on a silicon surface. The packing of the atoms is much less dense than within the crystal itself.

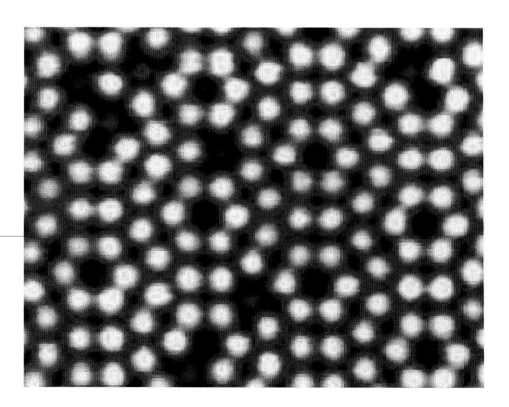

An arrangement of a finite number of differently shaped polygons that completely cover a plane with no gaps or overlaps is called "tiling the plane." The arrangements can be periodic or non-periodic. The non-periodic arrangement in this photograph is made up of only two polygon shapes and is an example of Penrose tiling. Such tiling arrangements form the basis of physicists' understanding of crystalline-like substances called quasicrystals.

Because matter has wavelike properties, waves can be used to shape the fabrication of atomic-scale structures. In this image, produced by the National Institute of Standards and Technology in 1995, scientists placed individual chromium atoms on the surface of a silicon crystal. To control the spacing of the chromium atoms, the researchers directed a standing wave of laser light across the silicon sheet. Standing waves, such as those that form in piano strings or organ pipes, have alternating regions of high and low energies. Spots with the lowest energy are called nodes and occur every half-wavelength. As the chromium atoms were deposited on the surface, they naturally settled into nodes in the laser wave. As a result, the space between each atom is 213 billionths of a meter—half the wavelength of the laser light.

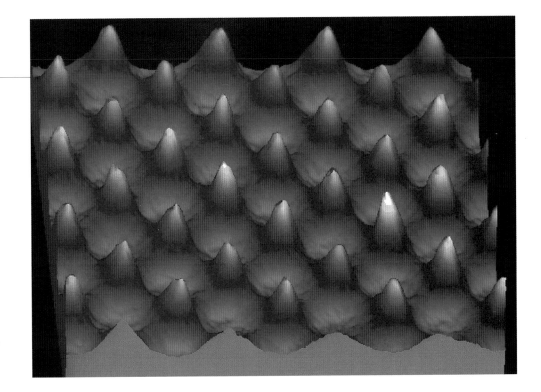

Superconductivity

Meanwhile, other physicists were wrestling with a baffling solid-state effect first seen in 1911. Around the turn of the century, there was a brisk international competition to see who could get closest to absolute zero—the point at which there are no lower temperatures—and how various materials behaved when they got there.

Hydrogen had been liquefied in 1898 at 12 degrees Kelvin. (The Kelvin scale uses familiar Centigrade degrees, but starts the numbering at absolute zero. Thus 0 K is -273 C.) By 1900, liquefication of the recently discovered element helium had become the new goal. Calculations suggested that it could not occur above 6 K.

Dutch physicist Heike Kamerlingh Onnes finally achieved it at 4.2 K and subsequently used liquid helium to cool numerous other elements and observe how their properties changed with temperature. In doing so, he discovered to his astonishment that mercury abruptly lost all resistance to electric current below a "critical" temperature. It did not simply become a much better conductor. In a sudden "phase transition" of the sort that occurs when liquid water solidifies, it apparently ceased having any resistance at all! It was a "superconductor."

If a current was initiated in a frigid ring of such material, and the electrodes then removed, the current continued to circulate unimpeded indefinitely. Eventually other elements, including lead, tin, zinc, and aluminum were also shown to be "superconductors" at ultra-low temperatures.

Of course, physicists had long known that the resistance of metals decreased with temperature. But there was no conceivable way that they should lose *all* resistance. Indeed, there was even reason to believe the opposite: near absolute zero, some had predicted,

This photograph, taken around 1988, shows the phenomenon of stable magnetic levitation. A lozenge-shaped permanent magnet is held aloft above a disk of superconducting materials. Superconductors, as the name suggests, offer no resistance to electrical current. But they have another related property, called the Meissner effect: superconductors actively expel magnetic fields by means of circulatory currents on their surface. These surface currents exactly cancel out the external magnetic field inside the superconductor. Outside the superconductor, the magnetic field created by the circulatory surface currents repels the external magnet causing it to levitate above the superconductor. Transportation engineers hope to exploit the effect to hold railway cars suspended above their "tracks," thus eliminating friction.

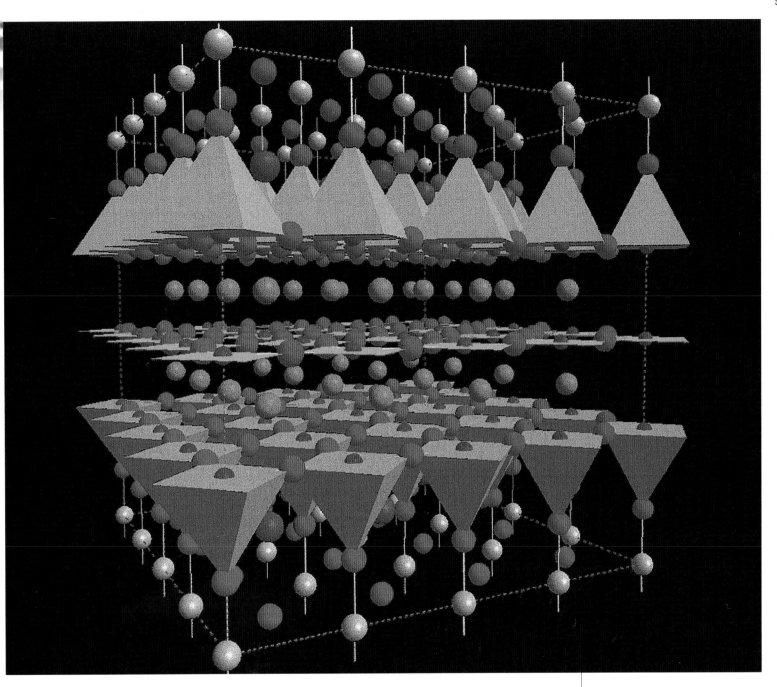

Dutch physicist Heike Kammerlingh Onnes discovered superconductivity in 1911 at temperatures only a few degrees above absolute zero. Since then, researchers have tried to find or invent materials that will become superconducting in much more normal conditions. They are still far from the goal of a room-temperature superconductor, but have gradually raised the threshold. Shown here is a schematic representation of a crystal composed of mercury, barium, calcium, copper, and oxygen that becomes superconducting at -140 C.

electrons would have so little energy that they would stay bound to their atoms, preventing current flow and giving the material a large, if not infinite resistance. Experimentalists and theorists leapt to explore and explain the superconductor phenomenon, which soon displayed even weirder anomalies.

In 1933 German-born physicist Walther Meissner and colleagues demonstrated that superconductors expel magnetic fields. That is, when exposed to a magnetic field, a superconducting material develops small eddy currents near its surface whose magnetic fields exactly cancel out those of the external field. The Meissner effect is so robust that a magnet will actually "levitate" above a superconductor, and engineers are now working on designs for high-speed trains that will use the effect as a frictionless means of support.

Physicists would later find another class of superconductors that permitted a few magnetic field lines to penetrate, but trapped them in place, permitting the superconducting electrons to move around the threadlike "pinned" field lines. On the other hand, if the field was strong enough, it would destroy the superconductivity of either kind of material.

By the mid-1930s, experimentalists had accumulated mystifying observations galore. Yet there was still no suitable theoretical explanation. The German-born brothers Heinz and Fritz London, who had determined that the eddy currents act within a few millionths of a centimeter of the surface, made some early progress in comprehending what quantum-mechanical processes might be at work. But it would take twenty more years of intense effort to devise such a theory. And when Americans John Bardeen (of transistor fame), Leon Cooper, and J. Robert Schrieffer finally did so in 1957, the resulting "BCS" explanation was every bit as strange as the phenomenon itself.

At sufficiently low temperatures, they reasoned, electrons in superconductors formed into matched doublets called Cooper pairs. They work something like this: as the first electron in a pair approaches two positively charged atoms—actually ions, since some of their electrons are off in circulation—within the metal lattice, the electron's negative charge pulls the ions slightly closer together. After the first electron passes, the ions begin to relax back to their original position. But before they can do so completely, the second electron in the pair comes along and takes advantage of the altered configuration, which exerts an attraction on the second electron, allowing it to move effortlessly through the lattice.

(That does not mean, however, that the two electrons in a Cooper pair are physically close to one another. In fact, they may be thousands of atoms apart. They form a quantum pair not because of proximity, but because they are perfectly matched in two ways: they have exactly opposite momenta, traveling in different directions at the same speed, and they have opposite spin.)

As billions of such pairs propagate along the superconducting material, they act much like the total superposition of wave functions seen in Bose-Einstein condensates (described in Chapter 3) that can occur when particles or atoms occupy exactly the same quantum condition.

Were this not bizarre enough, the science of superconductors would change again before the end of the century. After BCS, scientists around the world were busily investigating the onset of superconductivity, often with an eye to whether it might occur at higher temperatures. The early superconducting wires could be used to make many devices; for instance, currents flowing through superconducting coils can create stupendously powerful magnetic fields. Cooling them with liquid helium, however, was highly expensive and very difficult. So except for specialty uses, such as constructing high-field magnets for MRI scanners, superconductors remained generally impractical. What engineers wanted was a material that would superconduct above 77 K. That's the temperature of liquid nitrogen, which is easy to handle and cheap as beer. Prospects were dim: BCS theory put a ceiling of 30 or 40 K on the phenomenon.

Warming Up

By the early 1980s, many new superconducting substances had been found. But none worked above 23 K—still far too low a temperature to achieve conveniently and at low cost.

The answer to this problem came by what might appear an unlikely route. Swiss physicist K. Alex Müller of the IBM research center in Zurich had become intrigued by the possibilities of oxides. They seemed utterly inappropriate: most oxides were classic insulators because highly reactive oxygen atoms tend to snatch up nearby electrons, making them unavailable for conduction.

But Müller had noticed an odd thing about aluminum. By itself, it became superconducting at 1.1 K; when mixed with grains of aluminum oxide, however, the temperature rises to 2.2 K. Perhaps the oxide somehow helped raise the critical point. For years, he and IBM colleague Johann Bednorz tested dozens of different combinations, with invariably disappointing results.

Finally, they hit upon a compound of barium, lanthanum, copper, and oxygen previously described by French chemists in another context. The IBM researchers found that its resistance disappeared at 35 K—50 percent higher than the world record and right at the edge of what BCS theory allowed.

They announced their results in 1986, kicking off a global scramble for even higher temperatures. Within two years, other labs had tested compounds containing obscure elements such as yttrium, and temperatures reached 90 K and then 125 K. As the twenty-first century dawns, however, scientists are still far from understanding how these "high temperature" superconductors work.

But the transistor and superconductor research blitz, and the growing understanding of how quantum-mechanical processes affect atomic bonds, lattice structure, and other properties, taught physicists a vast amount about how to custom-create solid-state materials. This has allowed transistors to become smaller and smaller, so that it is now possible to put millions of them on a single chip of silicon. In the next twenty years, it is expected that transistors will become so small that quantum effects will become important and entirely new techniques will be needed by the electronics industry.

Matter by Design

The practical consequences of condensed-matter and materials science in the last one-third of the twentieth century are ubiquitous and literally too numerous to mention. The ability to manipulate atomic bonds and precisely alter crystal structure has made the walls of metal and plastic containers stronger and thinner; replaced steel with lightweight polymers such as Teflon and Kevlar; created high-speed transistors from unusual composite semiconductors such as gallium nitride; and produced solid-state diode lasers no larger than a grain of sand.

Physicists devised various ways to build up precision-tailored solids by depositing a stream of atoms and molecules layer by layer, constructing a new generation of "super-alloys." Using techniques such as plasma spray, they made possible a range of coatings for products ranging from razor blades to artificial hip joints, military armor, and aircraft surfaces. A related method called molecular beam epitaxy allows researchers to "grow" flawless and meticulously controlled materials from scratch. And physicists can now move individual atoms on material surfaces and study the effects of the atomic placement.

Various micro-procedures have led to fabrication of gears and other mechanical devices half the thickness of a human hair, and to manufacture of crystals so small that they are invisible and can be used in sunscreen. Electron-doping techniques have produced cylindrical arrays of carbon atoms—called nanotubes—that may ultimately function as submicroscopic wires.

One of the most spectacular applications of condensed-matter physics can be seen in every flat display screen and laptop computer: liquid crystals. In these strange entities, molecules can be made to line up with each other in response to subtle shifts in electric or magnetic fields. When their orientations change, so do their optical properties, from transparent to opaque and back again. That behavior is used to create characters and shapes on flat panels.

But perhaps the most illuminating aspect of liquid crystals, as the French physicist Pierre-Gilles de Gennes determined a couple of decades ago, is that they are typical of an entire class of physical changes that take place when other kinds of physical systems—such as superconductors, polymers, and magnets—pass from disordered to ordered states. At century's end, the exploration of such "phase transition" phenomena has become an intensely exciting field of research.

In 1955, physicists at General Electric created synthetic diamond for the first time. In order to simulate the conditions that cause natural diamonds to form at enormous heat and pressure in volcanic vents, the researchers compressed carbon at a million pounds per square inch and raised its temperature to 1,370 C in the presence of a metal catalyst. The catalyst helps the carbon atoms rearrange themselves into patterns of stiff, 3-dimensional bonds that make diamond the hardest naturally occurring substance. The result, shown here, is a synthetic diamond of about 1 carat, still wearing a dusting of catalyst.

As the twentieth century progressed, physicists grew increasingly sophisticated in their ability to construct microscopic arrangements of molecules and atoms. One highly successful method is called chemical vapor deposition, in which a fine mist of compounds is made to settle on and adhere to a surface. This image shows crystals of cesium iodide, used in an X-ray screen, that have been deposited on gauze by physicists at Royal Philips Electronics in the Netherlands. The herringbone pattern, which follows the weave of the gauze fibers, inhibits side-to-side scattering of light.

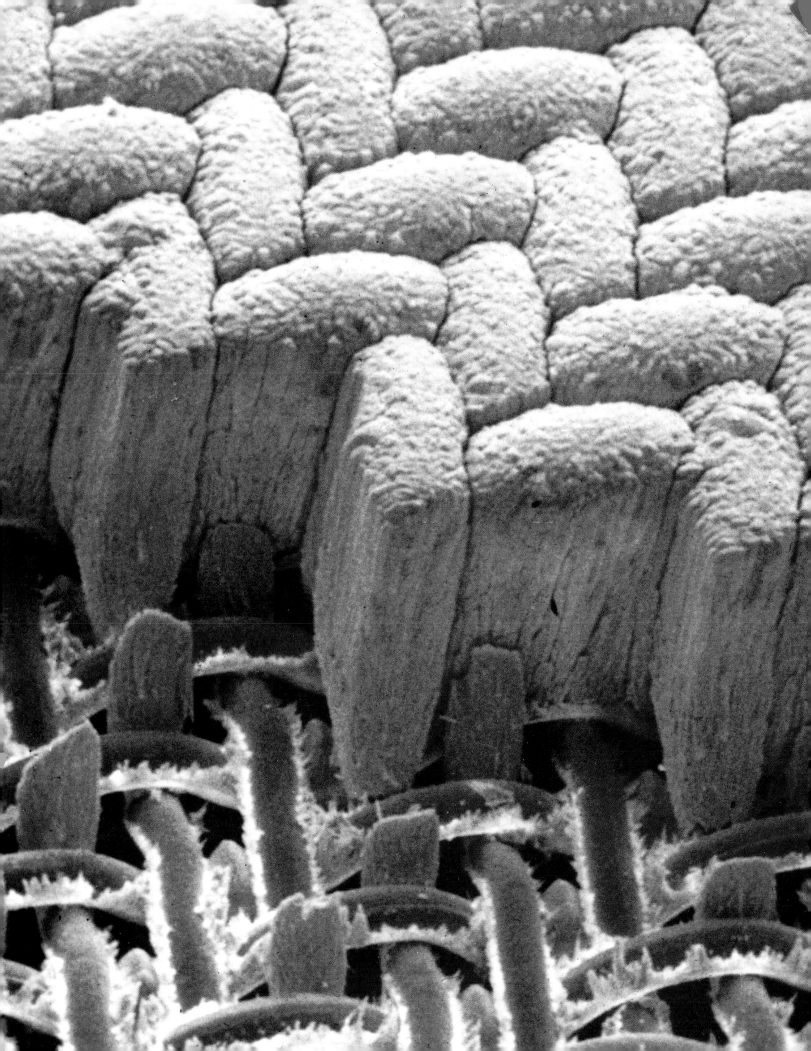

An STM image reveals a 4μ-thick film of gallium antimonide grown by a method called molecular beam epitaxy on a substrate of gallium arsenide. Spiral-like structures grow around dislocations in the film. The ability to determine and control material structure at the atomic level will revolutionize the next generation of electronic and optoelectronic devices.

Sometimes important materials don't have all the desired properties. For example, the atomic spacing in the crystalline ceramic material lanthanum aluminate is very close to that in some of the new, high-temperature superconductors. This means lanthanum aluminate would make an excellent base on which to deposit superconducting films. Unfortunately, good crystals of lanthanum aluminate are very hard to grow: this image shows the stair-like imperfections that often form. Material scientists are trying to find ways to reliably grow crystals without these imperfections.

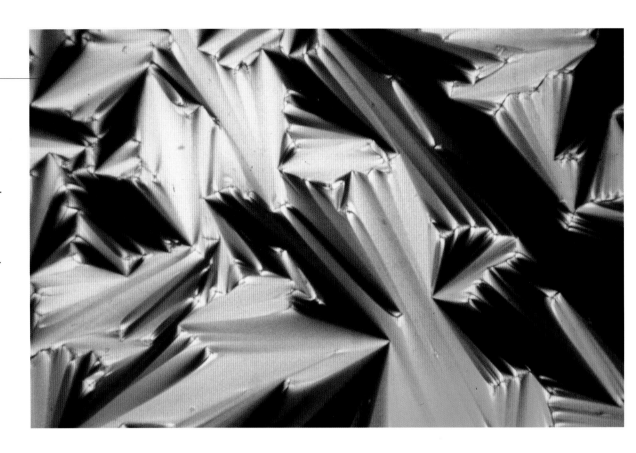

Liquid crystals have become the display mechanism of choice for a host of devices from pocket calculators and digital watches to portable computer screens. They consist of long-chain molecules that align themselves with certain preferred orientations. As this image shows, groups of these molecules can have nearly crystalline structures, which means that they can transmit, bend, or block light depending on their molecular orientation. And they are practically liquid, which means that the molecules can move around in their environment.

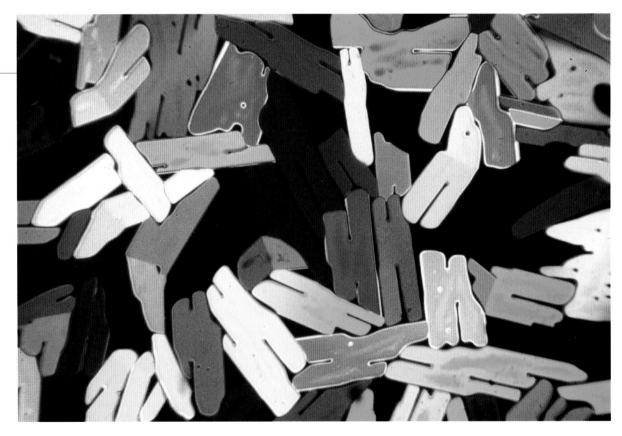

Physicists found that by using tiny electric fields to change the alignment of liquid crystals such as these relative to a light source, they could make them either opaque or transparent. The resulting pattern of light and dark spots makes numbers and letters on electronic device displays.

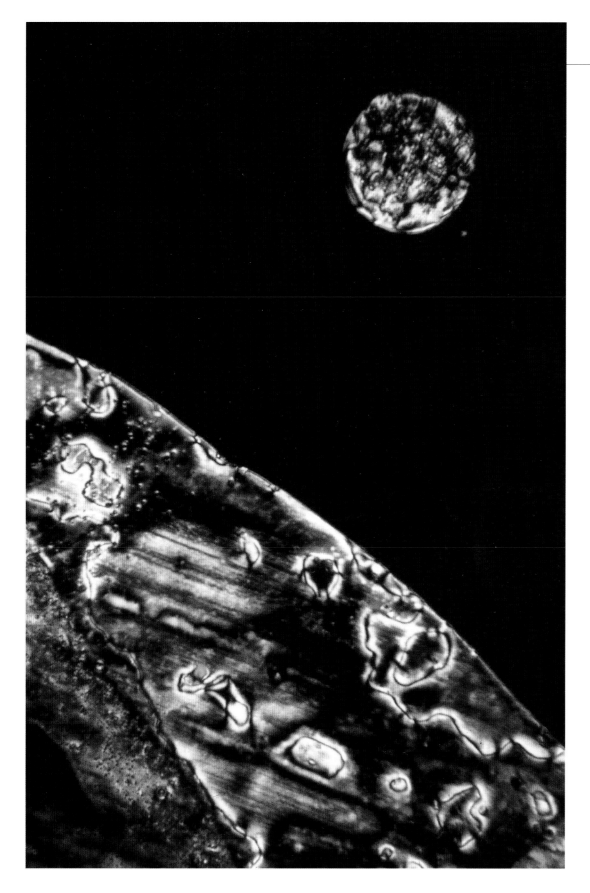

One form of a liquid crystal is called nematic. In this particular liquid crystal form, the long axes of the crystal molecules maintain a parallel or nearly parallel arrangement to each other. This picture shows the initial formation of a small nematic liquid crystal above a much larger one. Nematic crystals are used for displays on wristwatches, calculators, and panel meters.

Because the dimensional scale of micro-devices now reaches into the billionths of meters, the construction of such structures is called nanotechnology. Using extremely fine focusing techniques on a sheet of special compounds, scientists can use patterns of projected light to create tiny shapes. Wherever the light strikes the compound, it alters the chemical properties; areas that remain in shadow stay the way they were. The sheet is then bathed in a solution that dissolves the exposed areas. That technique, called photo-lithography, can be used to produce parts for incredibly small devices. One such micromachine is shown right alongside a grain of pollen (top right) and coagulated blood cells (top left and bottom right). The scale bar is 50 millionths of meter wide—slightly less than the width of a human hair. The gear array shown below has been driven at speeds up to 240,000 rpm—nearly one hundred times faster than a car's crankshaft is turning while speeding down the interstate. Micromachines may eventually perform a host of functions, including medical procedures inside small blood vessels.

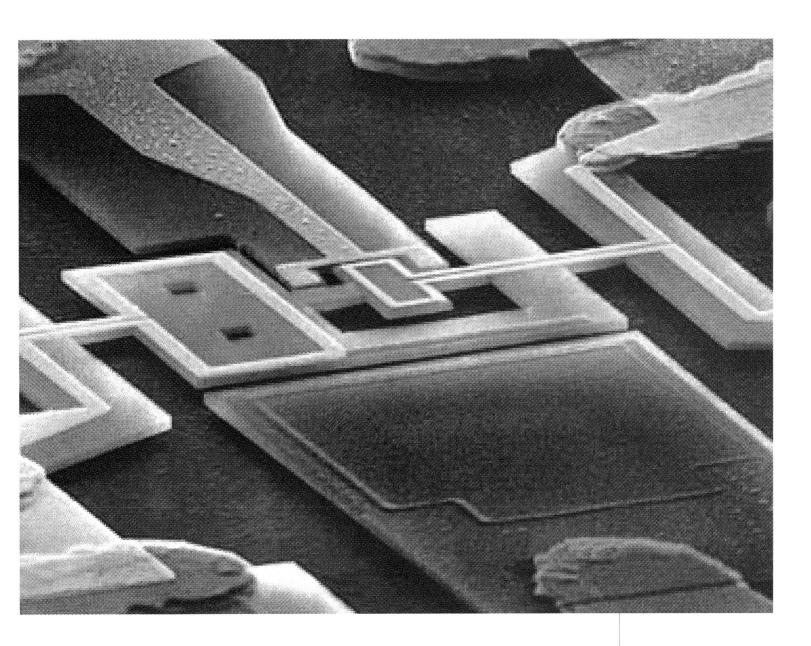

The apparatus pictured here, built into a single crystal of silicon, is a sensor capable of detecting extremely small electric charges—smaller, in fact, than the charge of a single electron. The electrodes are plated in gold.

Science now recognizes four fundamental forces in nature: gravity, electromagnetism, and the "strong" and "weak" forces that affect subatomic particles. Well into the twentieth century, however, only the first two were known. Fully half the forces that shape the physical world were discovered and explained in the past few decades! Moreover, despite the pioneering research of Thomson, Rutherford, and others, until midcentury physicists were aware of only a small fraction of the basic subatomic components of matter that today are familiar concepts to schoolchildren.

The effort to find, comprehend, and control those forces and particles ranks among the preeminent accomplishments of the human intellect. It made possible the exploitation of nuclear fission to generate enormous amounts of electricity—including 20 percent of America's and 70 percent of Western Europe's. It led to the development of nuclear weapons with destructive capabilities unprecedented in the history of war. It revealed the seething alchemy that powers the sun. And it is still enabling physicists to peer into the very heart of matter, where the deepest mysteries lurk at dimensions so small they approach nothing at all.

The Neutron

This busy image was recorded at CERN, the European Laboratory for Particle Physics, in Geneva, Switzerland. Physicists accelerate protons around a ring that is 656 feet in diameter and then propel them into solid targets. The collision produces a spray of smaller particles whose tracks are revealed in CERN's liquid hydrogen bubble chamber. Magnetic fields cause charged particles to curve, creating the spiral bubbletrail shapes seen here. The direction of the curvature indicates whether the particle has a positive or negative charge, and the degree of curvature indicates the mass and/or speed: lighter and slower particles curve more; heavier and faster particles less.

By the 1920s, many questions about the nature of atoms had been answered. Protons and electrons had been identified, their masses determined, and their equal but opposite charges ascertained to good accuracy. Atomic structure, though still puzzling, was tentatively explicable: because atoms are electrically neutral, there must be one proton for every electron. But the known masses of elements made it clear that there were far more nuclear particles than could be offset by the number of orbiting electrons. So into the early 1930s, it was assumed that the nucleus contained enough additional electrons to balance the books.

That model, however, was badly at odds with several observations and theoretical dictates. Although no one knew what to put in its place, Rutherford had come to suspect that there was another, unknown type of nucleon, probably some kind of electron-proton composite.

Evidence for this strange hypothetical entity would emerge from the worldwide scramble to understand the process called radioactivity that had been discovered around the turn of the twentieth century. Rutherford and Frederick Soddy, a chemist at McGill University in Montreal, after investigating several substances such as thorium, arrived at a critical insight: radioactivity was not so much a property of certain elements as the by-product of a disintegration process whereby one element might become another.

The idea that some nuclei were inherently unstable contradicted the prevailing dogma that elements were immutable. But by 1910, it was established beyond doubt. Although the source of nuclear instability would not be understood for decades, its effects were more immediately useful.

By studying the rate at which one element is transformed into another (which Rutherford expressed in "half-life": the time it takes for half the atoms to break down into their "daughter" products), scientists found an accurate method of dating very old rock samples and organic materials. Improved over the course of the century, radioactive dating is now an indispensable tool in geology, anthropology, and paleontology.

Physicists recognized three types of radioactive emissions. In alpha radiation, atoms eject a massive entity that in 1908 was found to be identical to a helium nucleus. In beta radiation, they give off energetic electrons—thus apparently confirming the notion that there were, in fact, electrons in the nucleus. And in gamma radiation, they emit high-energy photons.

Some laboratories, though, were detecting a form of induced radiation that was presumed to be some kind of gamma ray, but didn't quite act like anything then known. For example, the French wife-husband team of Irène Curie and Frédéric Joliot showed that when they bombarded beryllium with alpha particles, something came flying out of the metal that would knock protons out of paraffin.

British physicist James Chadwick tested that effect on different light elements and determined that the "rays" were actually massive objects capable of producing recoil protons. Chadwick was able to show that the peculiar entity striking the proton had a mass nearly indistinguishable from that of the proton itself. Yet it could pass through as much as 8 inches of lead, whereas a proton at the same velocity is stopped by as little as 1/4 mm of lead. Because electrical charge reduces penetrating power, Chadwick reasoned, the particle (or, as he imagined it at the time, the bound proton-electron pair) must have a very small, or even zero, charge. In 1932, he published his findings.

The neutron, as it was called, was Rutherford's long-expected neutral component, needed to explain the mass of the nucleus. By the mid-1930s it had been confirmed as a single particle, finally eliminating the theoretical notion that there were electrons in the nucleus. Later it would become an immensely valuable probe of solid matter: a beam of neutrons can be used somewhat like X-rays to discern the geometric structures of crystals. Although they have no charge, neutrons do have a magnetic moment, and they "feel" magnetic fields while passing through the crystal. As a result, their exit pattern reveals the arrangement of magnetic ions in the sample.

In addition, physicists can irradiate atoms with neutrons to make artificial radioactive isotopes such as iodine and phosphorus for use in cancer therapy and tissue testing. More recently, scientists have learned how to use a process called neutron "activation" to detect trace constituents as small as trillionths of a gram by bombarding the sample with slow-moving neutrons. The atoms in the sample capture the neutrons, forming unstable isotopes. As they return to stable conditions, they shed energy in the form of gamma rays. The frequency of the gamma photon emitted is distinctively different for each element, allowing a highly sensitive and specific assay of ingredients.

The Neutrino

The discovery of the neutron had by no means exhausted the nuclear puzzle. Another enigma was the nature of beta decay, which did not seem to follow the rules for conservation of energy. In the other two types of radioactivity, physicists could account for the energy content of the particles and photons. If a substance emitted an alpha particle with less than its maximum possible energy, it also gave off a photon that contained the remainder. The sum of the two was always the same; energy was conserved. So, as expected, were electrical charge and momentum. But in beta radiation, atoms of the same element emitted electrons in a very broad range of energies with no accompanying photons. Either beta decay violated one of nature's most dependably invariable laws, or there was something else going on.

In 1931, Wolfgang Pauli proposed that there must be an elusive particle that was emitted along with the electron and carried the missing energy. Italian physicist Enrico Fermi went even further. In 1934, he argued that there must be a "weak" force (weaker, that is, than whatever force binds particles together in the nucleus) that was responsible for nuclear decay by prompting a metamorphosis in particles. It could cause a proton and an electron to merge into a neutron; or, conversely, it could prompt a neutron to decay into a proton, an electron, and something he called a "neutrino." The name is a diminutive in Italian, as well it might be: Fermi figured that it must have a tiny or zero mass and no electrical charge. When he submitted a paper on the concept to the prestigious journal *Nature*, the editors turned him down. The idea was just too weird.

And so, it seemed, was the supposed particle. By its very nature, it would be almost impossible to detect. In fact, it took more than twenty years until two American physicists, Frederick Reines and Clyde Cowan, devised a suitable experiment. Theory predicted that if a certain kind of neutrino hit a proton, it would convert the proton into a neutron and release a positron—the antimatter counterpart of the electron. When the positron was annihilated by contact with a neighboring electron, it should produce two photons with a precisely predictable energy.

Moreover, the newly created neutron would have a predictable velocity, causing it to slam into a nearby atom and emit a third photon with a characteristic energy content. In 1956, Reines and Cowan finally saw the unmistakable three-photon "signature" they were looking for in a liquid-filled detector. The neutrino was real. So was the bizarre, but now undeniable, weak force.

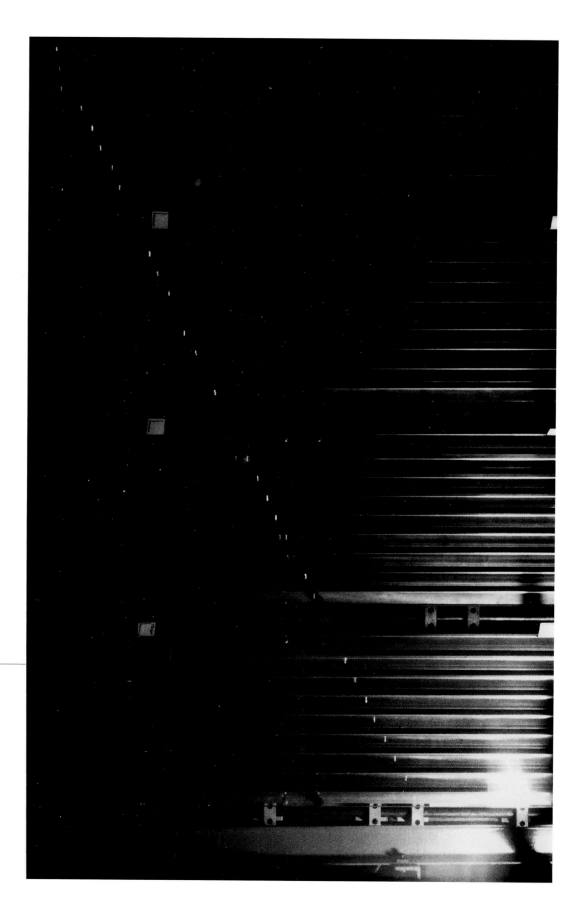

Some long-lived subatomic particles, such as the electron or proton, are relatively easy to study directly. But other, more exotic species exist for only a trillionth of a trillionth of a second before they break down into other particles. Their fleeting lives are observed only in sensitive detectors. This 1962 image, from Columbia University, shows the path of a particular type of cosmic ray called a muon. The series of experiments in which this image was taken revealed that there were at least two kinds of neutrinos—the elusive, nearly massless particles that interact very weakly with matter.

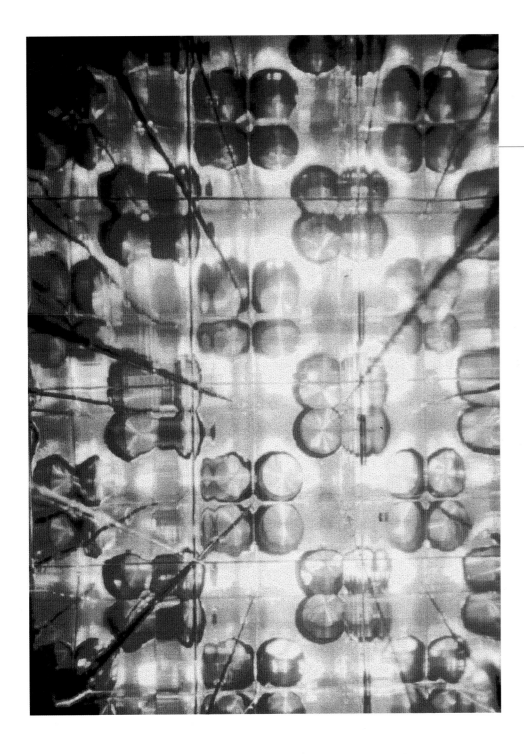

Neutrinos are difficult to detect, but a precise knowledge of their properties is necessary for understanding on both the smallest (fundamental particles) and largest (cosmology) scales. The Karmen experiment, a joint United Kingdom–German project, uses the neutrino detector shown in this picture. The experiments are being carried out at the Rutherford Appleton Laboratory in England.

Nowadays, experiments can be conducted with neutrino beams. But even today, neutrinos are horrendously difficult to observe. They interact so weakly with matter that most could shoot right through several million miles of lead. Some physicists, however, hope to use exactly that ability to "X-ray" the Earth by detecting subtle changes that occur in carefully prepared beams of neutrinos that are shot through the planet.

Fission

In addition to the discovery of the neutron, 1932 also witnessed the advent of the first practical particle accelerators. Physicists were no longer restricted to natural radioactive sources of particles. John Cockcroft and Ernest Walton converted lithium to helium by hitting it with highly accelerated protons, and many labs began looking at what happened when nuclei were bombarded with neutrons.

At the same time, a general theory was emerging that viewed the nucleus as a somewhat elastic spheroid. Like a drop of water, it could be deformed and set to oscillating by absorbing a particle, and was held together by the subatomic equivalent of surface tension.

In fact, the nucleus maintained a balance between opposing forces. One was simply electrical. After it became clear that nuclei were populated only by positively charged protons and neutral neutrons, a troublesome question remained: because like charges repel, why didn't the positive protons fly apart? Clearly something was pulling them together more strongly than electrostatic repulsion was pushing them outward. This other force binding the nuclear particles was responsible for the general stability of most nuclei, reaching its greatest stability in iron, with 26 protons and 30 neutrons.

However, as atoms became very large, with more than 200 particles in the nucleus, they tended to become notably less stable. Some could absorb a particle without catastrophic change. But in many, the mutual repulsion of dozens of protons was just marginally offset by binding forces. Additional energy from a particle collision could disturb the precarious nuclear balance of forces, causing radioactive emissions or leading to the formation of new, different nuclei.

Fermi and others probed such nuclear reactions for years, firing neutrons at uranium and identifying at least four radioactive by-products. A major breakthrough came at the end of 1938, when Germans Otto Hahn and Fritz Strassmann found that the elements remaining after neutron bombardment appeared to be considerably lighter than the uranium target. Austrian-born physicists Lise Meitner and Otto Frisch offered a radical explanation: each target nucleus had split apart into two smaller nuclei.

(Lise Meitner's profound contribution to the understanding of fission is not as well known as it might be. Of Jewish origin, she was forced to flee Berlin in July 1938 and could not participate directly in important experiments with Hahn and Strassman that year, although her analysis was crucial to the discovery. Nor did her name appear on an important paper published in Germany by Hahn and collaborators. Hahn went on to win the Nobel Prize in 1944 for his work in fission. Meitner was considered but passed over in 1946. "The chance that I might become your Nobel colleague," she wrote to Hahn, "is finally settled.")

One of the most revolutionary concepts envisioned by Albert Einstein is the equivalence of mass and energy. That is, under certain circumstances, mass can become energy and vice versa. The original large-scale experimental demonstration of that phenomenon occurred on December 2, 1942, at the University of Chicago. Physicists there produced the world's first instance of controlled nuclear fission in the device shown opposite left. Radioactive uranium oxide (visible as lumps around the topmost layer of the "nuclear pile") was interspersed with layers of graphite, a form of pure carbon, to buffer and control the reaction.

Nuclear reactors now provide 20 percent of America's electrical power as well as numerous particles used in medicine and research. In most reactor designs, the radioactive material is covered by water that carries away some of the heat of fission. The photo opposite right shows how various control and sensor cables extend through the pool and down to the reactor core.

Few physicists had ever even imagined such a thing; but it made a certain kind of sense. Physical systems naturally seek the lowest energy state they can achieve, which is why water runs downhill. For some large nuclei, it is energetically more attractive to divide into two smaller, more favorable arrangements, a process Frisch named "fission." The surplus energy is released in an explosive increase in the kinetic energy of the two new nuclei.

In its most common form, U-238, the uranium nucleus is stable and does not split. (Of course, as we have seen in previous chapters, quantum mechanics demands that there is some finite probability that the nucleus will be in every possible condition at any moment. It is this probability that facilitates spontaneous radioactive emissions.)

But one isotope, U-235, with the largest ratio of protons to neutrons, is prone to fission when struck by a slow-moving neutron. What happens, physicists determined, is that within a thousandth of a trillionth of a second, the uranium nucleus breaks in two, typically ejecting two neutrons. Those two can go on to strike other nuclei, causing yet more fission and more neutron release in a pattern called a chain reaction. Fermi and colleagues demonstrated this process in the first working nuclear reactor at the University of Chicago at the end of 1942.

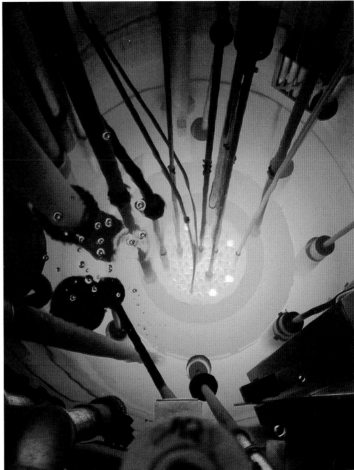

Nuclear fission, in the form of the "atomic bomb," helped bring an end to World War II. The top-secret development effort, called the Manhattan Project, was centered at Los Alamos, New Mexico and directed by J. Robert Oppenheimer, shown above. On July 16, 1945, Oppenheimer's team—which included some of the most famous American physicists of the twentieth century—succeeded in creating a blast that turned night into day over the New Mexico desert. The photographs opposite show the first seconds of the atomic era. Less than 6 milliseconds (ms) after the explosion (upper left), the fireball from the device, which was suspended about 100 feet above the ground, has not yet descended. The following three frames show the explosion at 90ms, 2 seconds, and 4 seconds, when the familiar mushroom cloud is about to form its stem.

Three years later, the super-secret Manhattan Project had harnessed the energy of fission in an "atomic bomb" that helped to end World War II. Scientists also discovered how to use the phenomenon to create even more fissile artificial elements such as plutonium, to produce radioactive isotopes of smaller atoms for medical and other uses, and to provide a controllable source of neutrons for material analysis and the rapidly expanding field of nuclear medicine. In 1956, the first fission-powered nuclear electrical plant went on line. Hundreds more would follow.

Fusion

Meanwhile, physicists had been struggling to understand other aspects of the nuclear world, especially the force that binds nucleons together. That interaction, first characterized by Japanese theorist Hideki Yukawa in 1935, is appropriately named the "strong" force: At the typical distance between nuclear particles, it is about 100 times stronger than the (quite powerful) electromagnetic force. But it only acts over a distance range of about a thousandth of a trillionth of a meter. The activity of the strong force is so complex that it took until 1974 for physicists to develop a suitable mathematical model to describe it.

Nonetheless, even early in the century researchers were aware of its effects, one of which is that the mass of a nucleus is measurably smaller than the masses of its individual component parts. By 1930, it was known that the mass of a helium nucleus is about 1 percent less than the aggregate masses of its constituents (then taken to be two protons and two electrons).

What happened to the missing mass? It was transformed to energy, according to Einstein's famous equation, $E = mc^2$, expressing the equivalence of mass and energy. That is, to disassemble a stable nucleus into its separate components, you would have to add considerable energy. Conversely, if you could squash those same components together hard enough to overcome their natural electrostatic repulsion, they would combine into a bound nucleus with less mass and release the difference as energy.

As early as 1920, British astrophysicist Arthur Eddington had suggested that such a combinatory process, fusing several hydrogen nuclei (that is, protons) into helium, might provide the source of the sun's energy. After all, about 70 percent of the sun was hydrogen, and most of the rest was helium. As it turned out, the proton fusion model agreed fairly well with the sun's energy output, but failed to explain the much greater luminosity of large stars, which seemed to be critically dependent on temperature. In 1938, German-born physicist Hans Bethe published a more comprehensive picture of stellar fusion, including reactions producing carbon, nitrogen, and oxygen, that was in good agreement with data from stars far larger and hotter than the sun.

As understanding of fusion improved, it too was subjugated to military use. Although no process on Earth could match the conditions at the core of the sun, where individual protons are fused into pairs, triplets, and foursomes, physicists found a way to produce fusion more readily by starting with the heavier isotopes of hydrogen, called deuterium and tritium, and then compressing and heating them with an atomic explosion.

The equivalence of mass and energy is important for nuclear fusion. Small nuclei of hydrogen fuse into larger helium nuclei that are slightly less massive than their components. The difference is converted to energy in the course of the reaction. Physicists hope to harness that power to generate electricity, and the process is being studied at sites such as the General Atomics DIII-D Takamak Fusion Facility in San Diego, opposite.

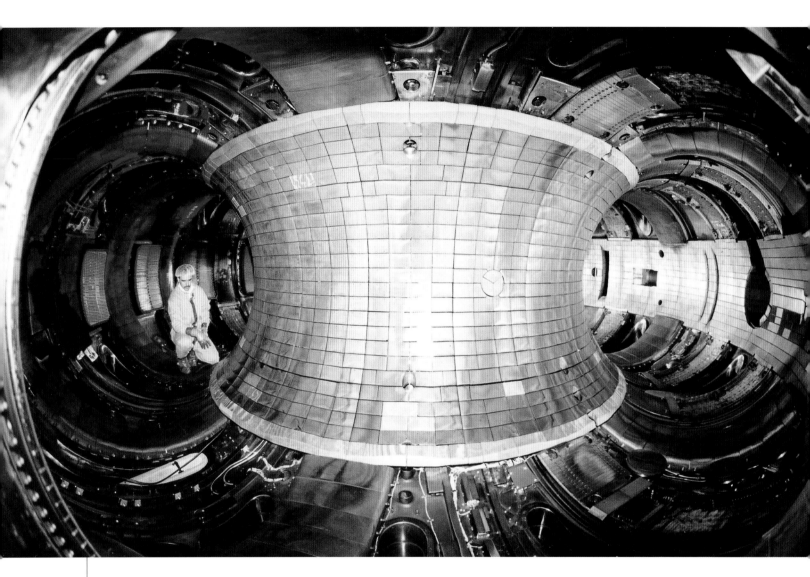

One way of inducing fusion is to create a plasma (a gas of particles heated to the point at which electrons separate from their atoms), hold it in place with powerful magnetic fields, and then add energy until the plasma "ignites." That is the function of this magnetic confinement chamber at the Princeton Plasma Physics Laboratory, where physicists trap a doughnut-shaped gas of hot hydrogen atoms in an enclosure covered with graphite-composite tiles.

This is an exterior view of the magnetic fusion reactor at Princeton. This unit, called the Takamak Fusion Test Reactor, is now being shut down. Research into magnetic confinement fusion is expected to continue at a more powerful international facility to be constructed early next century.

Another way to encourage nuclear fusion is to squeeze a small ball of hydrogen so hard that the nuclei bump and fuse—a method called inertial confinement. This time-exposure photograph, overleaf, from the fusion facility at Sandia National Laboratory in New Mexico shows electrical arcs that result when the device is fired. Enormous currents heat a cage of thread-like wires the size of a pill bottle, producing X-rays. The radiation vaporizes the surface of a BB-sized ball of hydrogen inside the cage, compressing it.

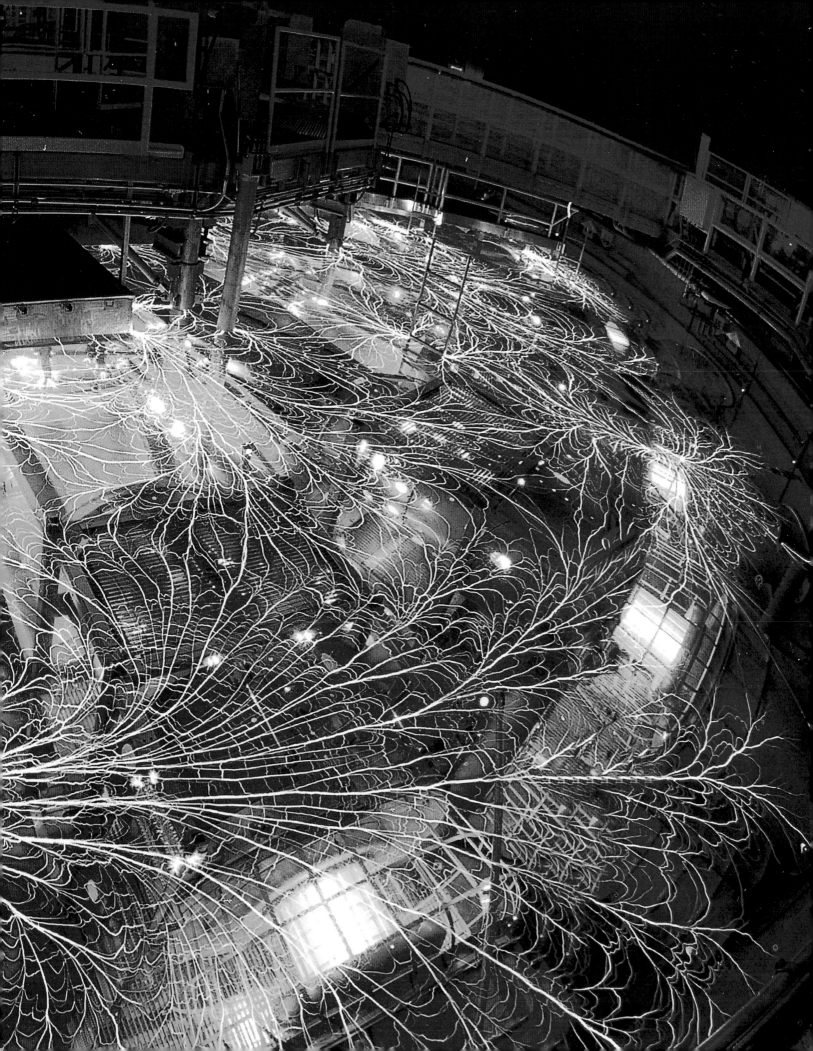

In order to accelerate charged particles to high energies, physicists give them repeated shoves with radio waves. Each perfectly timed shove boosts the particles' energy, just as pushing a child on a swing at the right moment will cause her to swing higher. The wave guides that direct the radio waves, such as this one from the Stanford Linear Accelerator Center, below, often have the spare beauty of Native American art.

In 1952, the first thermonuclear bomb was detonated in the Pacific, vaporizing a small island and setting off a worldwide fear of nuclear annihilation that persists to this day, despite international agreements sharply limiting such weapons.

At the same time, physicists began working to exploit fusion energy (gram for gram, about 60 times greater than the output of fission reactions) for peaceful purposes. Half a century later, that goal has not yet been achieved. But researchers have learned an extraordinary amount about trapping and heating hydrogen isotopes at energies high enough to strip the electrons from atoms, creating roiling plasmas in which nuclei get close enough to merge. Many continue to believe that a practical fusion power plant is feasible within decades.

Two methods have proven promising. In one, called magnetic confinement, the plasma is held by magnetic fields in a doughnut-shaped reactor chamber. In the other, inertial confinement, a tiny pellet of hydrogen isotope is compressed and heated by hitting it on all sides with high-energy laser beams or particles. So far, neither has achieved the "break-even" point at which fusion gives off as much energy as it took to create the plasma. Newly proposed facilities may do so.

The Particle Zoo

While one branch of physics was exploring fission and fusion reactions on a large scale, another was learning to study matter at the smallest dimensions from both natural and, later, artificial collisions of high-energy particles. The former are conveniently available in the form of cosmic rays, high-energy charged particles (usually protons) that reach Earth from space. Catching them and tracking their collision products, however, would tax the ingenuity of physicists, who were obliged to adapt or invent a number of detection devices.

One of the first was a version of the "cloud chamber" devised by British physicist C. T. R. Wilson early in the century. Used by Austrian physicist Victor Hess to study cosmic rays, and later by Carl Anderson to discover the positron, it was improved by English physicist Patrick Blackett. It employed a super-saturated mist that would condense into visible trails when any charged particle passed through it. Electric or magnetic fields in the chamber would bend the particles' trajectories in ways that would reveal their charge and mass.

In 1952, an improvement arrived in the form of the liquid-filled "bubble chamber" invented by American physicist Donald Glaser. It too caused particles to leave visible tracks as they caused the liquid to boil along their path. Still other scientists, notably English physicist Cecil Powell, adapted photographic plates that were carried aloft by balloons to record the tracks of particles in the upper atmosphere.

Cosmic ray research turned up an odd class of entities called mesons because their mass was midway between electrons and protons. One of the first detected initially appeared to be the force-carrying particles predicted by Yukawa's theory of the strong force. But that notion was dispelled as the properties of these new particles became better known. The Yukawa meson was discovered later, also in cosmic rays.

This is the heart of a 420-ton magnet designed for an experiment at Brookhaven National Laboratory's Alternating Gradient Synchrotron. The magnet is used in the search for rare decays of short-lived subatomic particles called kaons. This investigation is designed to test the theory of matter known as the Standard Model.

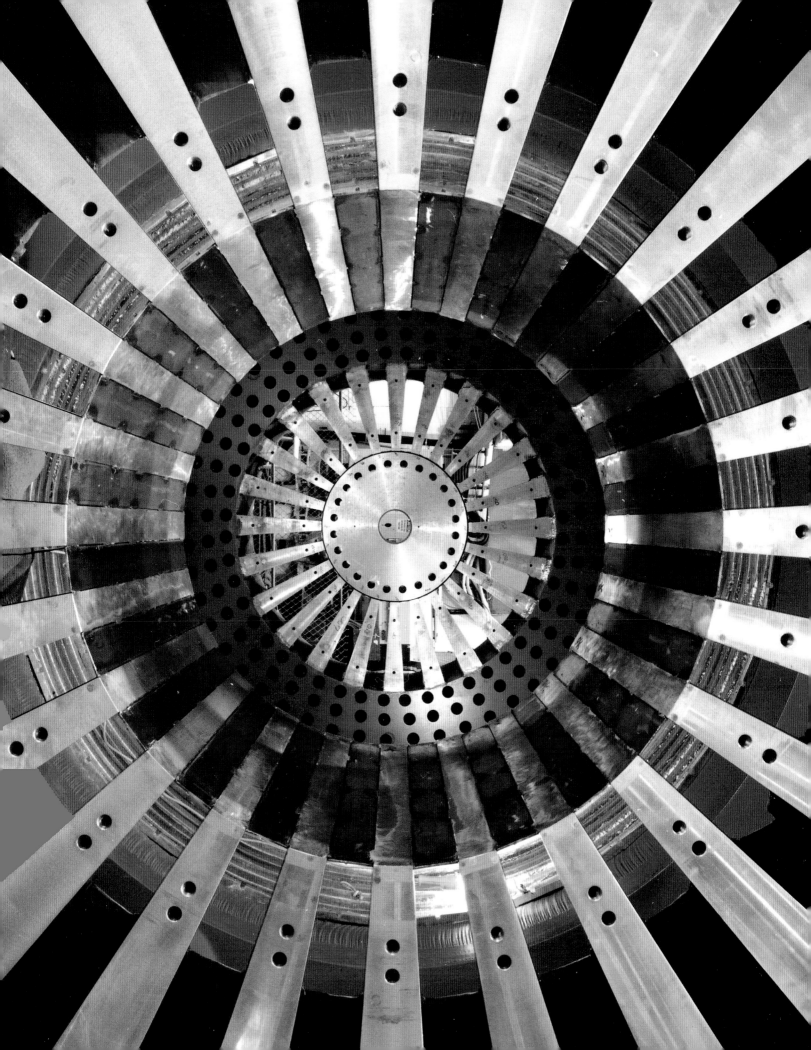

This spectrometer is used to study the highest energy elastic collisions between protons at the Brookhaven Alternating Gradient Synchrotron. Accelerated protons scatter from protons inside target nuclei and the resulting proton pair is analyzed by the spectrometer. A prediction of physical theory is that a special quark-gluon configuration of the proton can result from these collisions that, when analyzed, would reveal the interior protons of the target nucleus.

The larger and more energetic the objects that are smashed together in colliders, the greater the variety of particles created from the collision. Whereas the most familiar colliders accelerate beams of protons or electrons, a new device being tested at Brookhaven National Laboratory is designed to boost beams of massive atoms to nearly the speed of light around a 2.5-mile ring. Superconducting magnets in the Relativistic Heavy Ion Collider's beam line, shown here, will propel two opposing streams of elements such as gold into one another. The resulting spray of breakdown particles, physicists hope, will approximate conditions moments after the "Big Bang" that created the universe some 13 billion years ago.

The niobium superconducting cavity pair pictured here is the heart of the accelerator at the Thomas Jefferson National Accelerator Facility. Installed in cryostats, these devices accelerate electrons up to 4 million electron volts while bathed in supercold helium at -271 C or 2 K. A total of 338 cavities are installed at the facility, making it the largest installation of superconducting cavities in the world.

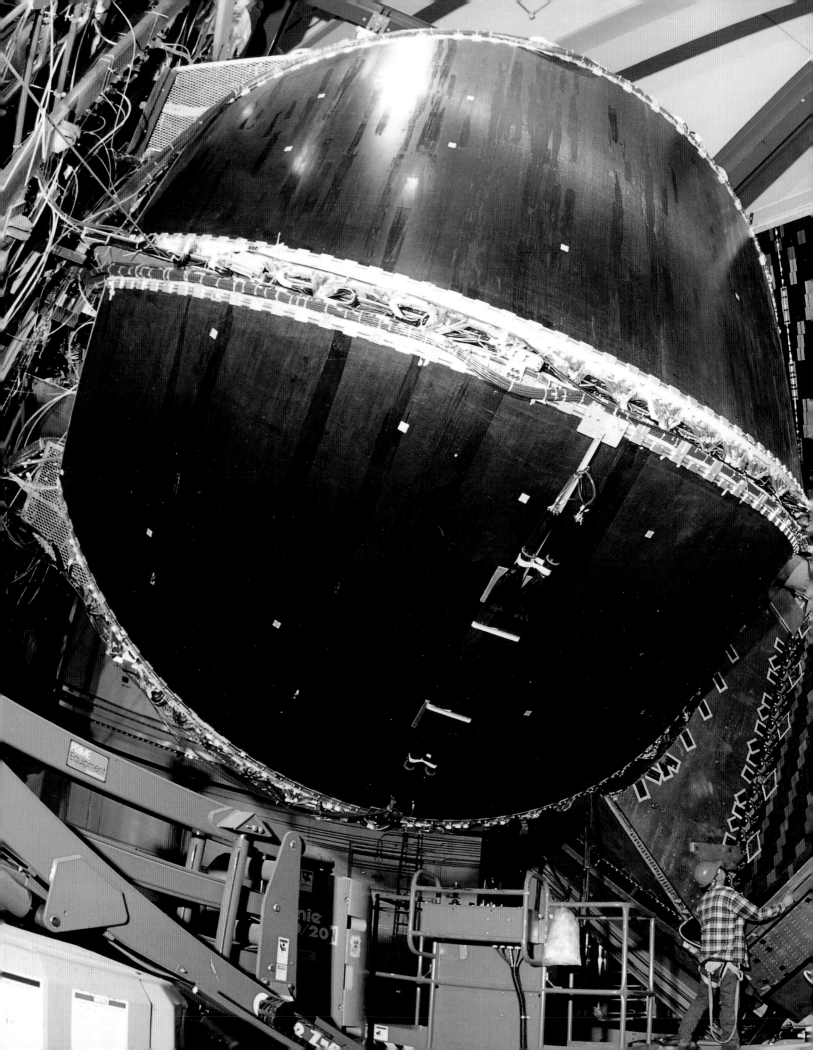

The proton-antiproton collider at the Fermi National Accelerator Laboratory outside Chicago is the world's most powerful. It has been the scene of many important discoveries, including the long-sought "Top Quark," the last and most massive of the quarks to be revealed. Pictured here is one of the two huge particle detectors, displaced out of its normal position for upgrading. The particle beam itself, which travels through the center of the circular opening, is less than 1/16 of an inch wide. The gigantic detector arrays are necessary to track the direction and energy of the various kinds of particles created in the collision.

This unique particle detector—the CEBAF (Continuous Electron Beam Acceleration Facility) Large Acceptance Spectrometer—was constructed over a seven-year period at the Thomas Jefferson National Accelerator Facility's Hall B. The spherical shape allows particles to be detected in many directions at once creating an incredible one terabyte of data a day to be analyzed. The black structure seen in the picture is the outside shape of six sections of wire chambers that surround the chamber holding the target of material with which the electron beam interacts. Scientists will use this detector to better understand the interactions between quarks and other particles, known as gluons, that hold quarks together to form protons and neutrons.

Throughout the twentieth century, physicists have studied the properties of gamma rays—the most highly energetic form of electromagnetic radiation. Traditionally, they have been examined by their collision effects on solid particles such as the components of atomic nuclei, or by their occasional spontaneous transformation in a particle-antiparticle pair. The gamma detector pictured here—called the Gammasphere and constructed at Lawrence Berkeley National Laboratory in California—provides a 100-fold increase in sensitivity over previous designs.

Cosmic rays were handy, but by nature uncontrollable. To study the phenomenon carefully, physicists wanted to be able to tune the exact energies and velocities of the colliding particles. That required the use of a new generation of particle accelerators (described in Chapter 1). Increasingly powerful accelerators began to generate a bewildering profusion of mesons and other even more mysterious particles.

In a high-energy collision, the resulting energy congealed instantly into matter (via $E = mc^2$), producing a host of previously unknown subatomic entities. The highest effective energies were achieved when energetic particles collided head-on with other energetic particles, providing the kinetic energy from both particles for new particle production.

In general, those collision products are not familiar particles such as electrons and protons that make up the everyday world. They are exotic, rapidly "decaying" entities that soon turn into numerous intermediate particles that in turn decay into a variety of simpler, more stable forms. Some last from a millionth to a trillionth of a second and can be observed directly.

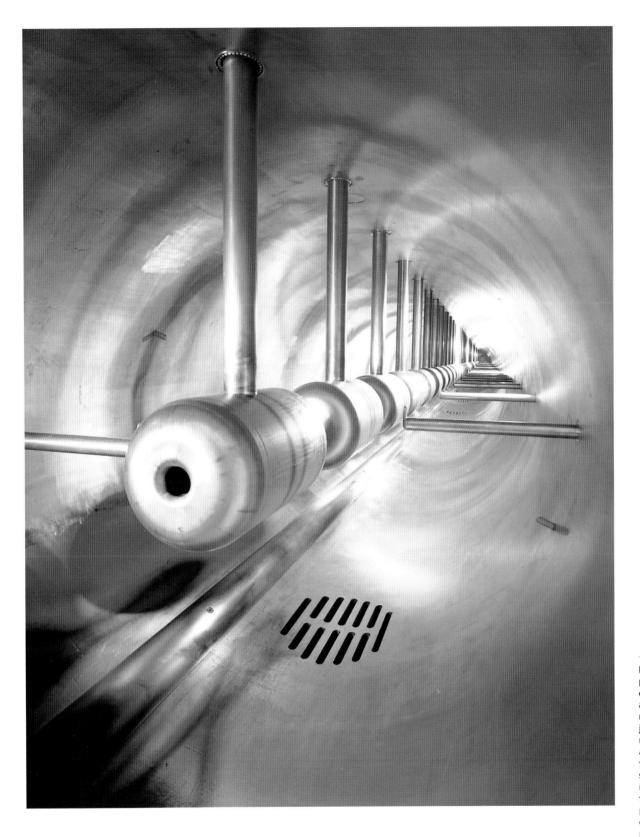

Because protons (and anti-protons) are charged particles, they can be accelerated and guided by electrical and magnetic fields. That is the job of the hundreds of magnets and radiation cavities along the Fermilab accelerator tunnel, part of which is shown here, that is 1.2 miles in diameter. Protons travel in one direction; antiprotons (the negatively charged anti-particles of protons) travel in the opposite direction. At the detectors, the two beams are steered so that they collide head-on.

Others have lifetimes around a trillionth of a trillionth of a second. Their nature can only be inferred by measuring the mass, energy, momentum, and electric charge of their numerous breakdown products, which researchers did with increasingly cunning and sensitive devices.

In a way, accelerator physics and detector science were too successful. By the 1960s, a hundred subatomic particles with different masses and charges had been observed. (The count is now up to 300.) Unless nature was more perversely complicated than anyone thought, there must be some underlying principles that explained the proliferating species of the particle zoo. To be sure, there were some helpful broad divisions. Physics recognized one fundamental category of lightweight particles, called leptons after the Greek word for "small," that included neutrinos, electrons, and a more massive electron-like particle called the muon that had initially been misclassified as a meson. Leptons are not affected by the strong force.

Another group was occupied by force-carrying particles called bosons after Satyendra Nath Bose: the photon, which carried the electromagnetic force, and four or five then-hypothetical entities that presumably conveyed gravity and the weak and strong forces.

But the most confusing abundance was in the category of particles collectively known as hadrons (which include protons and neutrons) from the Greek for "thick" or "bulky." In the mid-1960s, two California Institute of Technology physicists—Murray Gell-Mann and George Zweig—came up with a revolutionary idea. If protons, neutrons, and other hadrons were not truly elementary particles, but were made up of combinations of something even smaller, then the hadron inventory began to make beautiful mathematical and symmetrical sense.

Unfortunately, the theoretical particles—which Gell-Mann named quarks—would have to have properties that seemed radical if not outright ridiculous. For example, they were supposed to carry electrical charge in units of $1/3$ or $2/3$ the charge of the electron or proton. And they would have to embody properties so novel that physicists had to come up with new names such as "up" and "down" and "strangeness" to describe them.

The important contribution of the early quark model was its use as a basis for the classification scheme for the array of elementary particles. After the quark model was proposed, there were many attempts to find individual quarks, at accelerators, in the cosmic rays, and in the terrestrial environment—in mud, seawater, or anywhere. Not a quark was found. To most physicists, this was not surprising. Fractional charges were considered to be a strange and unacceptable concept, and the general point of view in 1966 was that quarks were most likely just mathematical representations, useful but not real.

In 1967, Jerome Friedman, Henry Kendall, Richard Taylor, and their collaborators began a series of experiments to study the structure of the proton at the Stanford Linear Accelerator Center (SLAC). Bombarding protons with electrons, they probed the inside of the proton, very much like using a powerful electron microscope. They found that there were pointlike constituents in the proton that had properties compatible with

The large ring on this photograph shows the location of the Large Hadron Collider at CERN, outside Geneva on the Swiss-French border. Particles speeding through the 17-mile underground accelerator will collide with more energy than is possible now even at Fermilab. The smaller circle indicates the position of a synchrotron that accelerates protons.

Fifty years ago, inspired by the bubbles rising in a glass of beer, University of Michigan physicist Donald Glaser had an idea: A particle speeding through a container of superheated liquid (that is, a liquid made very hot but kept from boiling in somewhat the same way that a kitchen pressure cooker works) would add just enough energy to the liquid to leave a track of boiling bubbles. Such a device, which he called a "bubble chamber," would be ideally suited to revealing the long trails of high-energy particles. In 1953, the first working bubble chamber obtained photographic images of particle tracks, pictured right. In 1960, Glaser received the Nobel Prize for his work.

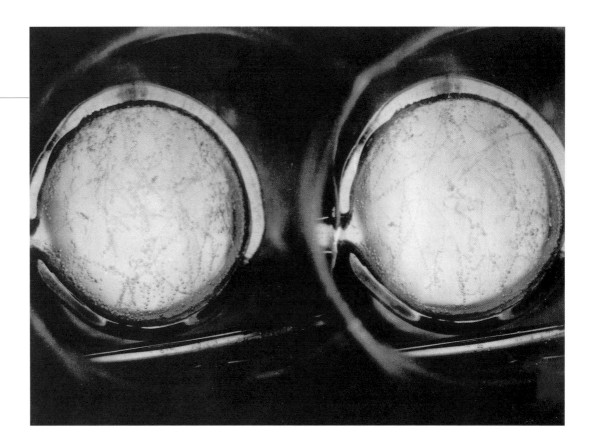

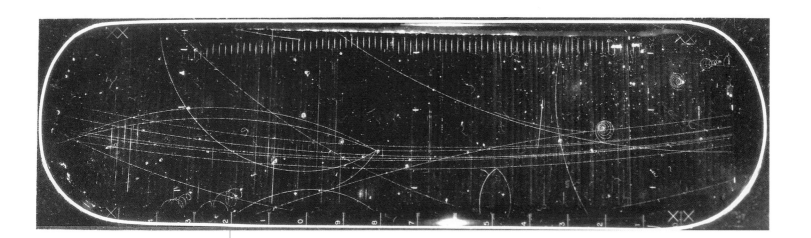

Donald Glaser's first bubble chamber was only a few inches in diameter. The one that produced this image, at Lawrence Berkeley National Laboratory in California, was six feet long. It could record extremely long tracks of the sort shown here.

Here an experimenter examines a photo from the fifteen-foot bubble chamber at Fermi National Laboratory. Bubble chambers have been replaced with fully automated electronic particle detectors. Computers can now rapidly scan through data collected and identify the few records of particularly interesting events. Then scientists can study the most revealing images.

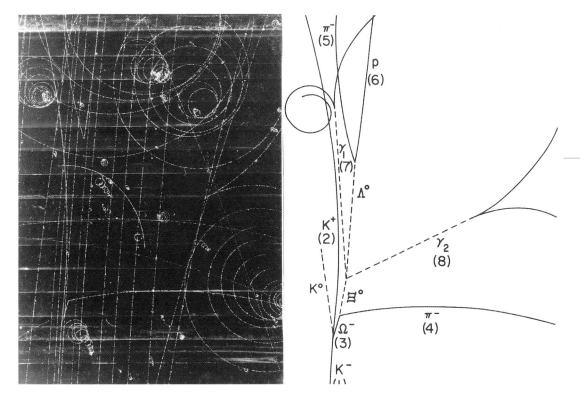

The ultimate goal of studying particle tracks is to see if they reveal unexpected physical events or interactions that are predicted by theory but have not been observed. This image is of the latter type. Recorded in 1964 with Brookhaven National Laboratory's 80-inch liquid-hydrogen bubble chamber—then the world's largest—it shows the production of a predicted particle called the Omega-minus (labeled Ω in the schematic diagram to the right of the photo). The observation of the Omega-minus and its decay products confirmed a key 1962 theory that contained the idea of quarks.

those of quarks. The combination of these results with those of subsequent experiments at CERN, the particle lab outside Geneva—which used neutrinos as the probe particles—unequivocally confirmed the fractional charges of quarks, or at least of the common "up" and "down" quarks that make up protons and neutrons.

But were there other kinds of quarks, as later predicted? For most physicists, doubt finally disappeared in 1974, when two rival groups independently found different kinds of evidence for the same particle—a configuration apparently including the "charm" quark predicted by American physicist Sheldon Glashow and others in the 1960s. Teams headed by Burton Richter at SLAC and Samuel Ting at Brookhaven National Laboratory on Long Island confirmed the existence of a particle subsequently shown to be a bound pair of charm and anticharm quarks.

That seemed to leave the particle roster with an extremely pleasing symmetry. There were two complete "families" of elementary particles, each with two leptons and two quarks. One included the familiar electron, its related neutrino, and the "up" and "down" quarks that in different combinations make up the protons and neutrons of ordinary matter. The other, more massive family—seen only in high-energy collisions—contained the muon and its associated neutrino, as well as the charm quark and its partner the "strange" quark.

But theory suggested that there should be three families. And in 1976, Martin Perl and others at SLAC detected an abnormally hefty lepton called the tau, an electron-like particle even more massive than the muon. Symmetry demanded that a third generation of leptons must have a corresponding third generation of quarks. But no one had ever seen such a thing.

Then in 1977, Leon Lederman and colleagues at Fermilab outside Chicago turned up evidence of yet another new quark about three times more massive than the one found by Richter and Ting. Since both theory and experiment showed that quark flavors come in related pairs, the discovery of this new "bottom" or "B" quark indicated that there must be a counterpart: the "top" quark that would complete the catalogue of elementary particles. It would, however, take nearly twenty years to find it. It would also take a landmark improvement in the power of particle colliders because, among other obstacles, it was nearly 40 times as massive as the B quark.

The more massive a particle one wants to observe, the more energy must be generated to create it. And by the late 1970s, physicists had reached extraordinary levels by crashing protons into protons, or electrons into positrons. But the mass of the top quark, as well as the masses of the particles that were presumed to carry the weak force, the W and Z bosons, were far beyond the capabilities of any facility available at the time. Much more energy would be needed.

To take the next step, researchers would have to collide protons with their antiparticles. Antiprotons had been discovered in 1955 by Berkeley scientists Emilio Segrè and Owen Chamberlain, and eventually scientists found ways to trap and accelerate them. The first proton-antiproton collider was built at CERN in 1981. By that time, particle projects had started to become what they are today: hugely complicated affairs involving collaborations of hundreds of physicists from several countries, along with even

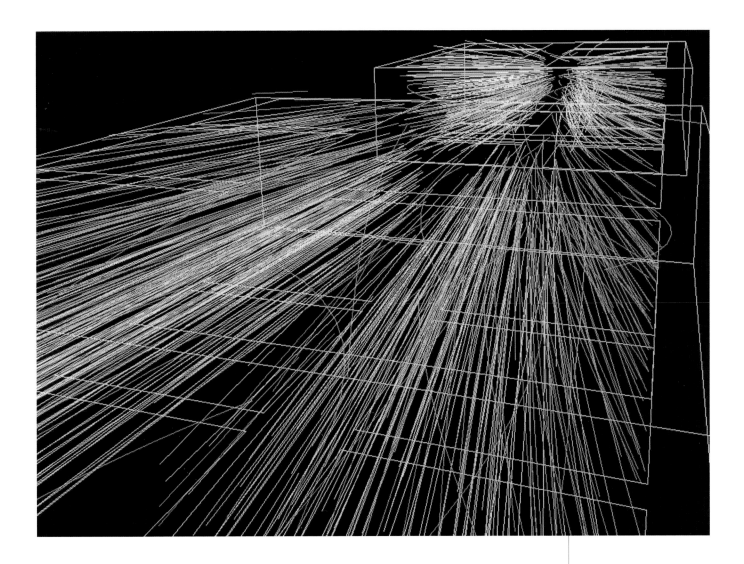

larger numbers of engineers and technicians, in which data are recorded from the solid-state equivalent of bubble chambers costing hundreds of millions of dollars each.

That investment began paying off almost immediately. In 1983, Italian physicist Carlo Rubbia and Dutch colleague Simon van der Meer and their collaborators found the W and Z particles responsible for nuclear disintegration. This came as a great relief to theorists, who had begun to distill all existing knowledge of forces and particles into a coordinated picture called the Standard Model.

Although certainly complex in its ramifications, the model is in many respects stunningly simple. It postulates that virtually every conceivable state of matter or energy can be embodied in fewer than twenty elementary entities: six kinds of leptons, six kinds of quarks, and four force-bearing bosons, including the gluon which conveys the strong force and also reacts to it. The Standard Model is the closest thing to a complete explanation of nature at the most fundamental level that humanity has ever devised. But like any theory, it is inherently provisional, subject to modification if and when experiments provide contradictory evidence.

These tracks are observed in a particle detector called a time projection chamber that is part of an experiment of an international collaboration at CERN. Particles resulting from the high-energy collision of two lead nuclei go through the gas of the time projection chamber. The resulting ionization is collected and digitized to obtain the above computer image. In this experiment, physicists hope to compress the nuclei to observe a new phase of matter—a quark-gluon plasma—thought to be the form of matter of the universe just after the Big Bang and in the cores of some neutron stars.

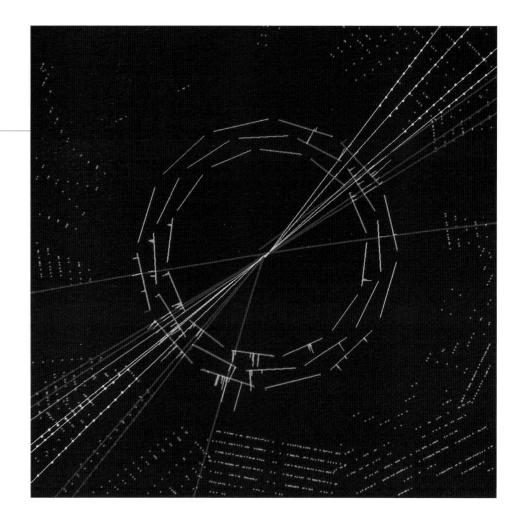

Physicists recognize particles produced in collisions by their electronic signatures, shown graphically by computers. The circle shows a computer-generated view of an experimental event that has a potential top quark signature. The particle tracks emerging from the center of a collision contain the particle's signature. This image came from the detector shown on page 141.

That is why there was so much global interest in the hunt for the top quark. It was nearly the last major constituent of the Standard Model that probably could be detected with contemporary equipment but had not been seen. Finally in 1995, two detectors at Fermilab—by then the world's most powerful proton-antiproton collider—found evidence of the top quark, with its utterly egregious mass thousands of times larger than the humble up and down quarks.

Thus the century that began with Rutherford's researchers squinting into microscopes in search of signs that the atom had a center ended with a comprehensive vision of particles 10,000 times smaller than the nucleus and the still-perplexing force that holds them together.

But profound questions remained. Why, for example, did the masses of the twelve leptons and quarks vary over at least six orders of magnitude? Why did nature seem to treat certain kinds of reactions unequally, a process called "symmetry breaking"? Where was the "graviton," the supposed boson that carried the gravitational force, and how could it be observed? At the close of the twentieth century, science is planning its assault on those formidable unknowns.

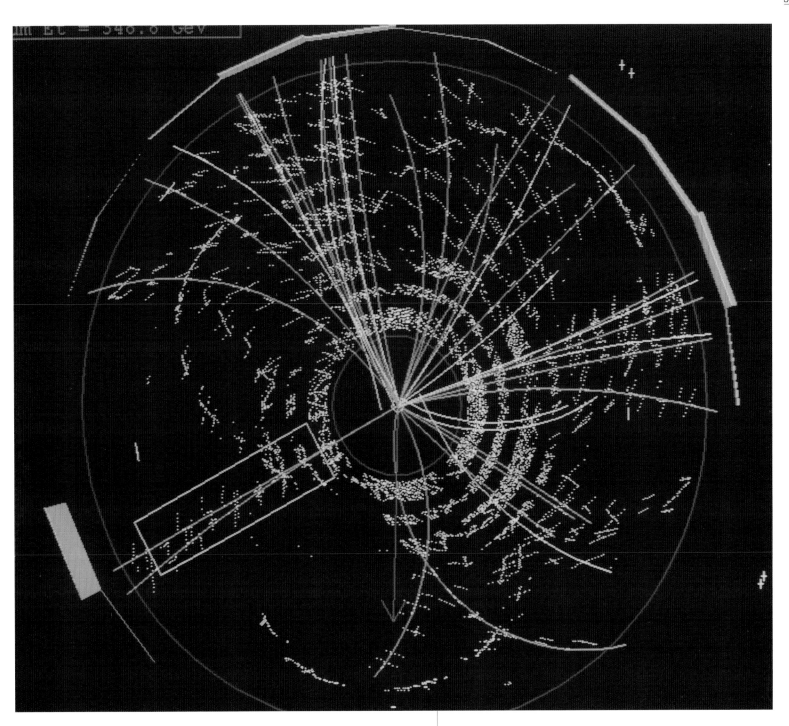

Computers are employed to scan through thousands of collision events that are recorded in electronic detectors arrayed around the collision site. When a key observation is identified, researchers can use the computer to produce a customized image of the event, with background "noise" and unimportant particle tracks removed. Graphic representations, such as this one from Fermilab, allow scientists to focus on the essential aspects of the physics.

6

CHAOS AND ORDER

By the last decades of the twentieth century, physics had probed the natural world in unprecedented scope and at scales ranging from the subatomic to the astronomical. Yet entire categories of readily visible, everyday phenomena remained stubbornly inexplicable.

Among the most problematic were certain kinds of physical systems with multiple parts in motion. And none was more infuriating than the apparently random behavior of moving fluids such as water or air. That kind of unpredictability seemed to make no sense: such systems are made up of individual macroscopic units—droplets or molecules—each of which is obliged to follow strictly deterministic Newtonian rules of force, motion, and position. Yet their *collective* properties often become unaccountably chaotic as they change over time.

For example, if you repeatedly throw the same rock into the air at the same angle with the same force, every time it will follow the same parabolic trajectory as it falls to earth. But pump the same volume of water past the same obstruction in the same pipe, and each time the stream will form a somewhat different and disorderly flow pattern, called turbulence.

That poses a paradox. Although every single water molecule is governed by inviolable laws, and its condition at any given moment is knowable in theory (at least within the uncertainty constraints of quantum mechanics), the aggregate motion remains unpredictable. Similarly, two days that begin with seemingly identical patterns of temperatures, pressures, wind direction, and vapor content in the air over a given region often evolve into two very disparate kinds of weather—as forecasters have learned to their frustration.

The seemingly chaotic nature of fluid turbulence makes it very difficult to study. One method is shown here. Scientists photographed a soap film as it fell vertically down a 6-foot-long tube. About midway in the process, the film passes a comb inserted into the column. Each tooth of the comb (visible as the round dark spots at the top of the image) acts somewhat like a wing, causing turbulence vortices to form in its wake. To visualize the turbulence clearly, the scientists lit their experiment with a sodium lamp beam that produces a brighter spot where the thickness of the film is greatest.

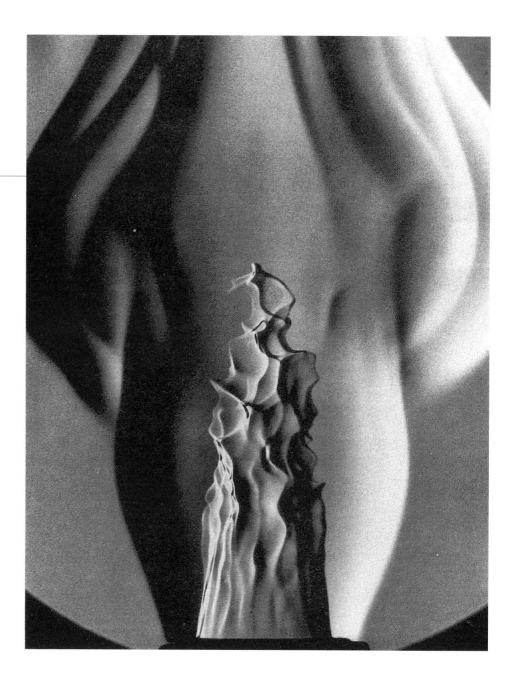

No everyday phenomenon is as difficult for physicists to model as turbulence. It pops up almost everywhere that smoothly flowing fluids encounter some irregularity. Suddenly an orderly stream of molecules seems to become chaotic, and there is as yet no satisfactory mathematical description of how the turbulent patterns emerge. This image shows how a thin sheet of burning propane and air, emerging from a circular source at the bottom of the picture, rapidly becomes wrinkled. The heat generated causes nearby air and gas molecules to form a chimney-like column surrounding the flame.

Classical physicists had tended to disregard that kind of messy behavior, preferring to concentrate on the aspects that were regular. They weren't ignoring the unpredictable so much as acknowledging that certain physical systems are governed by inherently "nonlinear" rules. That is, their output was not proportional to their input, so doubling the value of one variable doesn't produce twice the effect. It may produce five times the effect, or only one-third. Often, that's because the variables affect each other. For example, the rate of combustion depends on the speed at which reacting chemicals encounter one another; but the heat generated by the reaction speeds up the molecules, increasing the rate. Analogous processes make it extremely difficult to predict the motion of moving fluids such as air around an airplane wing.

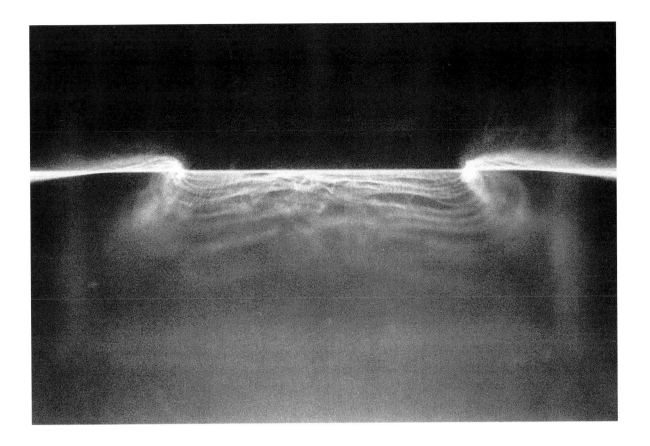

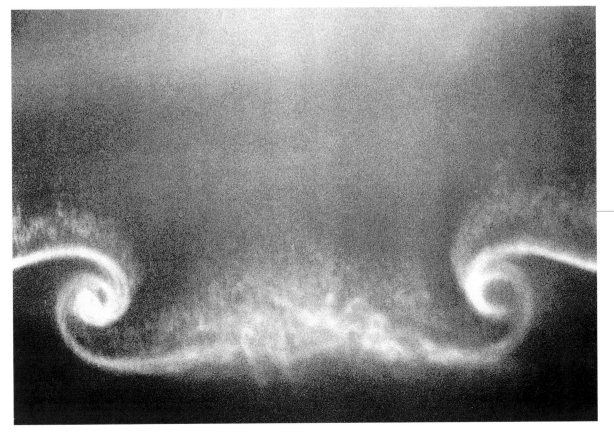

One source of the lift that enables airplanes to fly is the curvature of the wings. That shape causes air moving across the top of a wing to travel faster than air moving under it. The faster-moving air has a lower pressure, giving the plane its "lift." But when the airstreams above and below the plane's wing meet again, they produce a turbulence that is strong enough to endanger other aircraft flying behind. Often this turbulence takes a spiral or vortex pattern, as shown in this image. Here the upper image shows the flow near the wing, and the lower one shows the flow farther downstream.

Mysterious Motion

In fact, it had become increasingly clear since the nineteenth century that some aspects of nature were far less predictable than previously thought. Potential chaos even lurked in our own solar system, which in the classical Newtonian view was the very epitome of periodic regularity. French mathematical physicist Jules-Henri Poincaré had shown in the nineteenth century that it was practically impossible to predict the long-term motions of three or more astronomical bodies that influenced each other through mutual gravitational attraction. That is, even the stately orbits of the planets might eventually become highly irregular, given enough time.

That is not because the laws of gravity break down. Rather, it is because such systems are so exquisitely sensitive to very slight differences in initial conditions. Thus, even tiny variations in, say, the distance of closest approach between two planets as they orbit the sun will ultimately produce dramatically diverse orbital patterns as very small influences multiply over centuries.

Fluid dynamics was no more comprehensible. In the nineteenth century, physicists Claude Navier in France and George Stokes in Ireland had devised equations that described the properties of moving liquids. But in practice, it was usually impossible to solve them because the interactions were as enormously complicated as those for Poincaré's three-body problem. The eventual outcome was just as nonlinear and just as dependent on tiny variations in initial conditions. Even today, they are solved only as approximations with the use of supercomputers and remain a daunting challenge.

British physicist Horace Lamb voiced the general lament in 1932. "I am an old man now," he told a meeting of the British Association for the Advancement of Science, "and when I die and go to heaven there are two matters on which I hope for enlightenment. One is quantum electrodynamics, and the other is the turbulent motion of fluids. And about the former I am really rather optimistic."

By the early 1960s, however, help began to arrive in the timely confluence of several separate lines of research exploring the borderline between order and randomness. What investigators discovered over the next three decades, as they created the new science called "chaos" theory, was that the sharp Newtonian distinction between regularity and disorganization is illusory in two important ways.

For one, as Poincaré had observed, even the most fundamentally deterministic systems—that is, those whose rules can be described in relatively simple equations—can devolve into conditions so disorderly and unpredictable that they are essentially indistinguishable from what would be generated by chance. Conversely, however, many aspects of the natural world that *appear* chaotic may be understandable as the products of simple mathematical relationships—and thus perhaps describable and even controllable!

A flame can be regarded as a kind of feedback system. The burning creates heat, which alters the motion of the surrounding molecules, which in turn affects the shape of the flame, and so forth. The image opposite and the following two overleaf show the evolution of such a system photographed at intervals of 3/1,000 of a second. Note how spinning eddies, or vortices, are formed at the sides of the flame as hot rising gases encounter the cool, dense, quiet air surrounding the flame column.

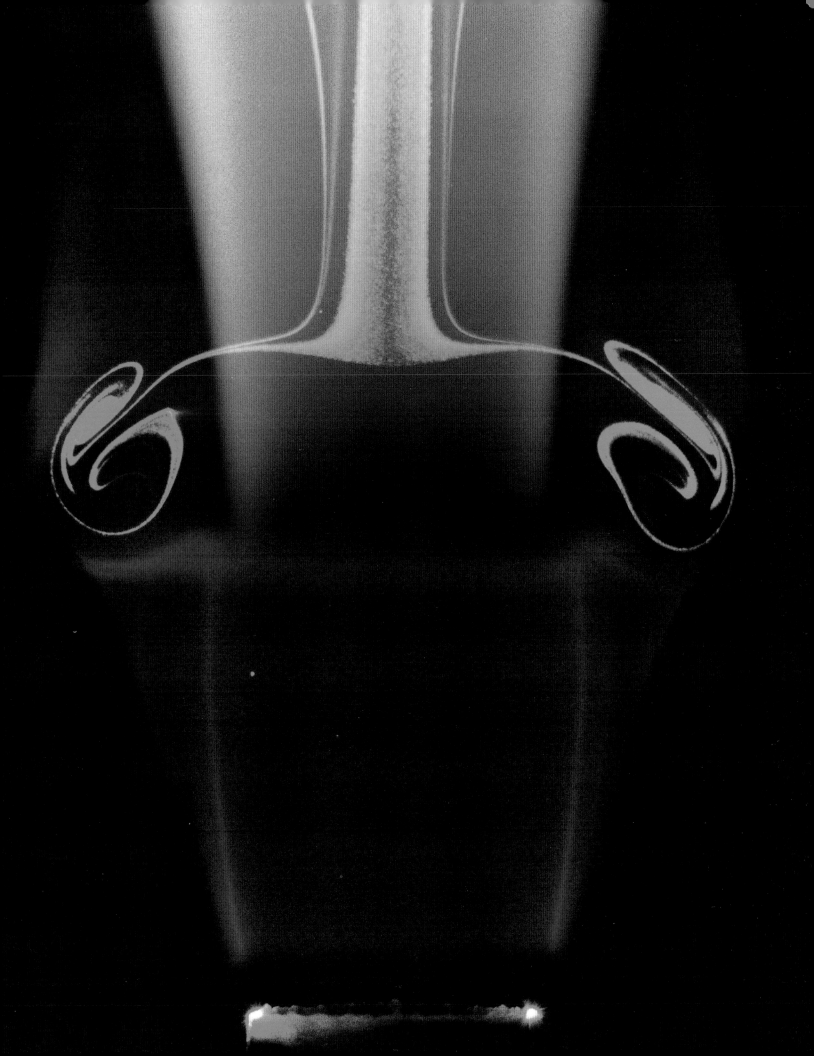

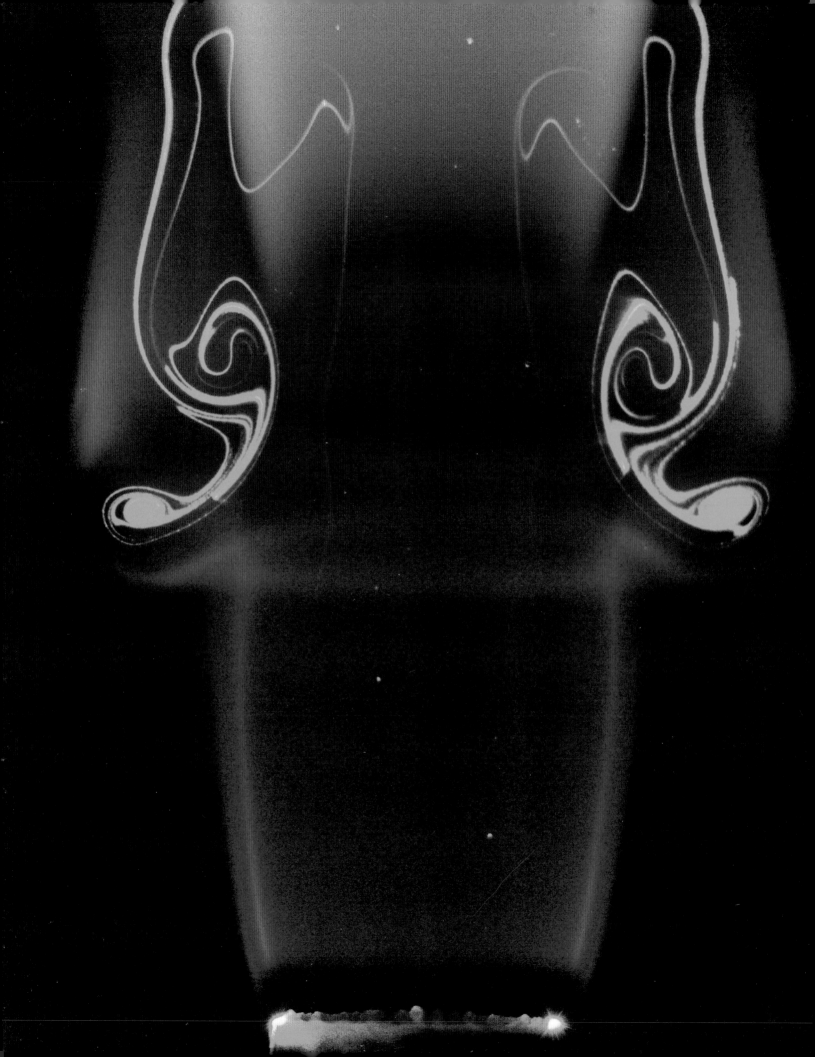

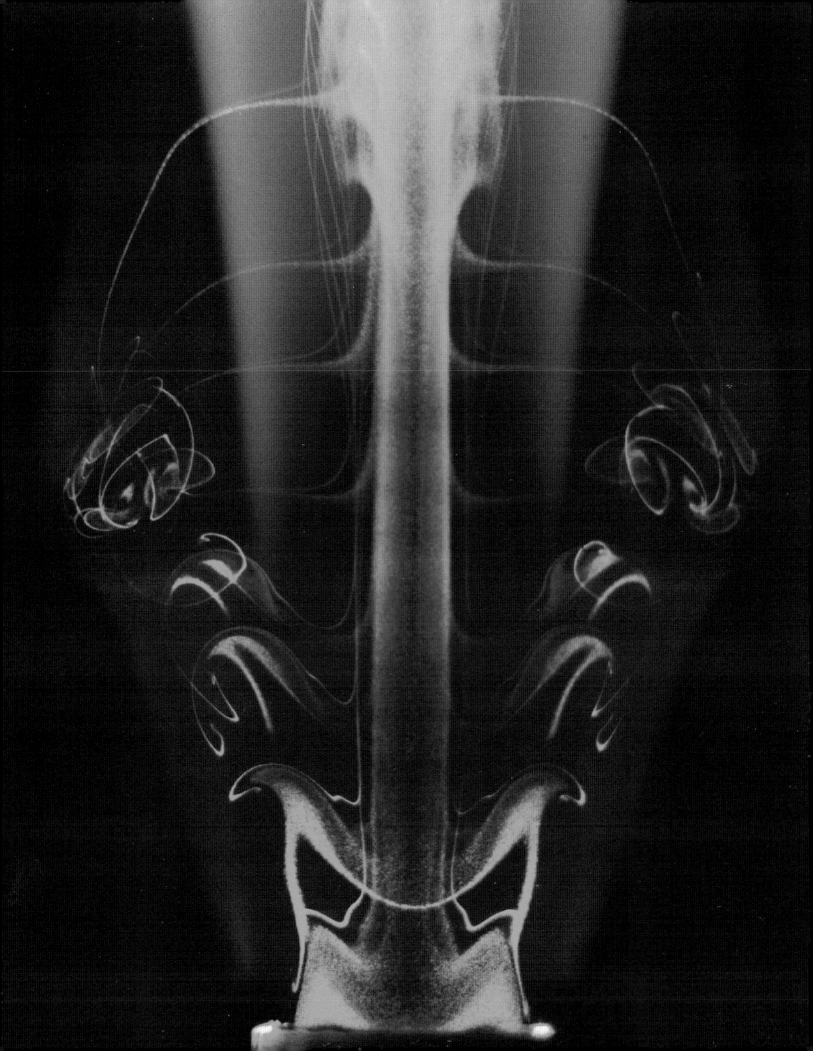

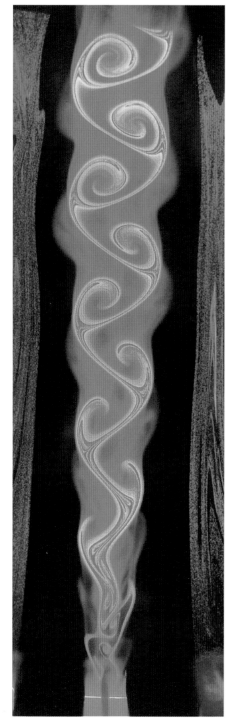
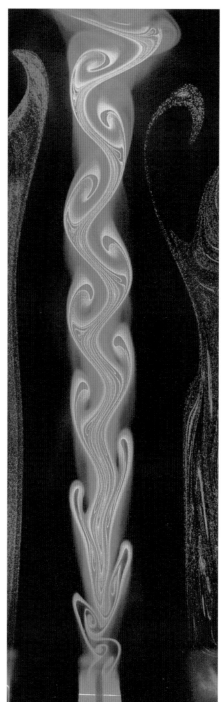
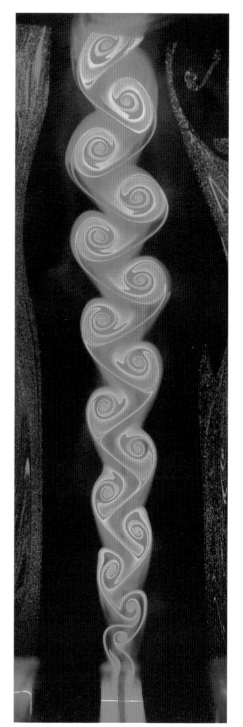

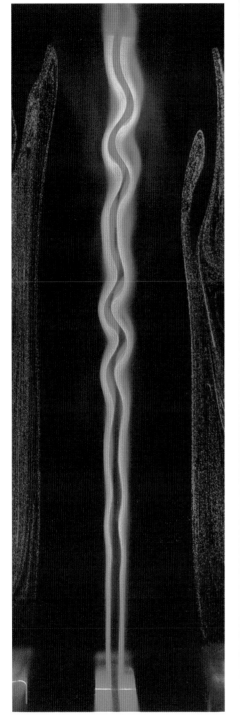
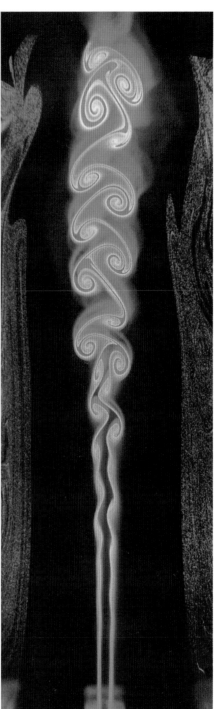
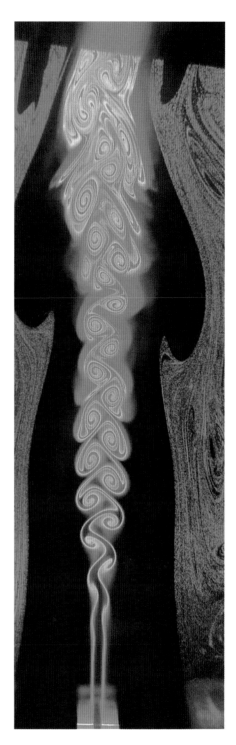

One of the frustrating—and fascinating—aspects of turbulence and of chaotic systems in general is that they often produce patterns that are midway between orderly and random. And although their final configurations are not predictable in a strict mathematical sense, there is a regularity to their structure. This series of images shows the patterns that are formed by changing the difference in speed between two streams of air. In all the photos, the first stream (shown as black) is moving around the outside of the rectangular object at the bottom of the picture. It flows at the same speed in all the images. A second stream (shown in green) is flowing from the slot in the rectangular object, at a different speed in each image. Spiral turbulence occurs where the two streams come in contact with each other, and the amplitude of the pattern is determined by the difference in speed. When the center (green) stream is moving either slower or faster than the outer (black) stream, the eddies are formed. In the fourth image, the two streams of air have about the same speed and no eddies are formed.

The Fractal World

One cluster of insights came from mathematics, where researchers were looking for ways to model the shapes of natural arrays such as coastlines, mountain ranges, plant growth patterns, and other configurations that do not embody simple Euclidean shapes. Chief among them was IBM mathematician Benoit Mandelbrot. He devised a new kind of geometry to describe these structures, which as he put it, "are irregular all over," yet "have the same degree of irregularity at all scales." He called such patterns "fractal," from the Latin word for broken stones, and began to find equations that would generate them.

By the 1960s, he had begun to see that the branching geometry of plants such as ferns or broccoli—in which each small part eerily resembles the whole in a condition called "self-similarity"—could be produced by certain kinds of "iterative" mathematical processes.

In those cases, the solution to an equation is fed back in as the new value for a key variable, and the calculation is made again. This gives the process the same kind of feedback effect seen in many apparently chaotic natural phenomena that also give rise to self-similar configurations. For example, watch the wake behind a moving boat or the way the rising smoke stream from a cigarette begins to curl. In both cases, the tiniest spiral patterns in the turbulence have the same shape as the largest swirls. Somehow nature imparts an overall regularity to their unpredictability.

IBM mathematician Benoit B. Mandelbrot first devised the geometry of fractals—objects such as coastlines or the edges of snowflakes, in which the smaller the scale on which the object is observed, the larger its perimeter appears to be. Many fractal objects embody "self-similarity." That is, the shapes of their largest units are repeated when the object is viewed at smaller and smaller dimensions. Perhaps the most famous fractal entities are Mandelbrot "sets" of solutions to certain equations. When the data are colored so that certain values get certain hues, the computer-generated solutions form shapes with an eerie fractal beauty.

Chaos in the Air

Similar events occur in weather patterns, where another set of insights into chaos emerged. In the early 1960s, meteorologist Edward Lorenz was attempting to create mathematical models of the way air currents move in the atmosphere. One might imagine that the overall processes of heat transport and exchange are so broad that small variations would tend to cancel themselves out or be subsumed by the larger pattern. Working with computerized simulations. Lorenz discovered just the opposite. He found that if he ran exactly the same simulation, but the initial conditions differed by as little as one part in 1,000, the outcomes were shockingly different as they evolved over time. The system was extraordinarily sensitive to tiny variations in its starting configuration.

The way a very small influence can magnify itself iteratively over time became known as the Butterfly Effect after the title of Lorenz's landmark paper, "Predictability: Does the Flap of a Butterfly's Wings in Brazil Set Off a Tornado in Texas?" Studying the process further, Lorenz calculated many hundreds of simulations and found that often the end results were not utterly random. Although never identical, they tended to cluster around certain typical values, just as the weather on a February day usually is of only two or three general types, rather than dozens.

This concept is called an "attractor": a point or set of values around which the outcomes of a nonlinear system cluster. Even though the results of an attractor-bound system never repeat themselves, they never become absolutely haphazard either. But some contain an eerily ominous feature: the attractor has two or more nodes. That is, the system has two or more general conditions into which it preferentially settles.

What this shows in theory is that a certain unpredictable combination of conditions can cause a system to jump rather suddenly between states. What it suggests in the real world is how an ostensibly very minor change in one or more variables such as sea-surface temperature, ocean currents, or wind direction may prompt the whole global climate system to shift between ice ages and interglacials. Analogous fluctuations in ecological parameters can shift animal populations with surprising speed between overpopulation and near-extinction.

Closing in on Chaos

By the 1970s, a decade after Lorenz and Mandelbrot conducted their early investigations (unbeknownst to each other), dozens of scientists had begun examining other kinds of attractor-driven nonlinear systems: whistling feedback in electronic circuits, the distribution of star motion in galaxies, the rise and fall of animal species, the changing tempo of water drops dripping from a tap, the spacing of mass in the asteroid belt, and the American stock market—to name only a few. Could there really be similar kinds of organizing patterns at work in such drastically diverse systems?

By the mid-1970s, thanks to the radically interdisciplinary approach taken by chaos enthusiasts, the answer was a tentative "yes." New insights were grafted onto more traditional research, such as the studies of turbulence by Andrei Kolmogorov and Lev Landau in the USSR, and work by David Ruelle and Michel Hénon in France, James

Many chaotic processes in the natural world, including weather and ocean currents, are driven by differences in density. The effects can also be studied in the laboratory. This image shows the penetration of a comparatively low-density fluid (purple and blue) as it rises through a denser fluid (yellow and orange), creating intense turbulence patterns. The green lines show areas in which the lighter fluid has stopped rising and the vertical velocity is near zero.

The same convection currents seen earlier in patterns of rising gases from a flame also control the shape and behavior of many natural systems, such as the moist air ascending to form storm clouds, or the motion of bubbling upward currents in a pot of boiling water. Shown here is a small-scale representation of convection patterns. A thin soap film is stretched between a hot surface at the bottom of the frame and a cold surface on top. The camera catches the motion of the molecules in the film over time. Starting at top left and proceeding down the left column and then the right, the film shows how the initially orderly structure is subjected to increasing turbulence as convective forces cause the warmer, low-level material to rise. By the end of the sequence, at bottom right, the configuration is extremely chaotic.

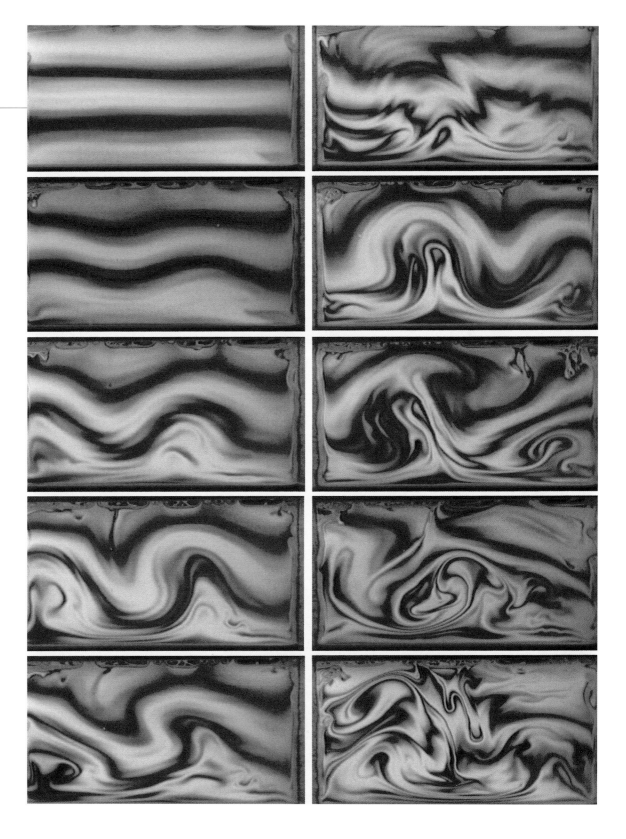

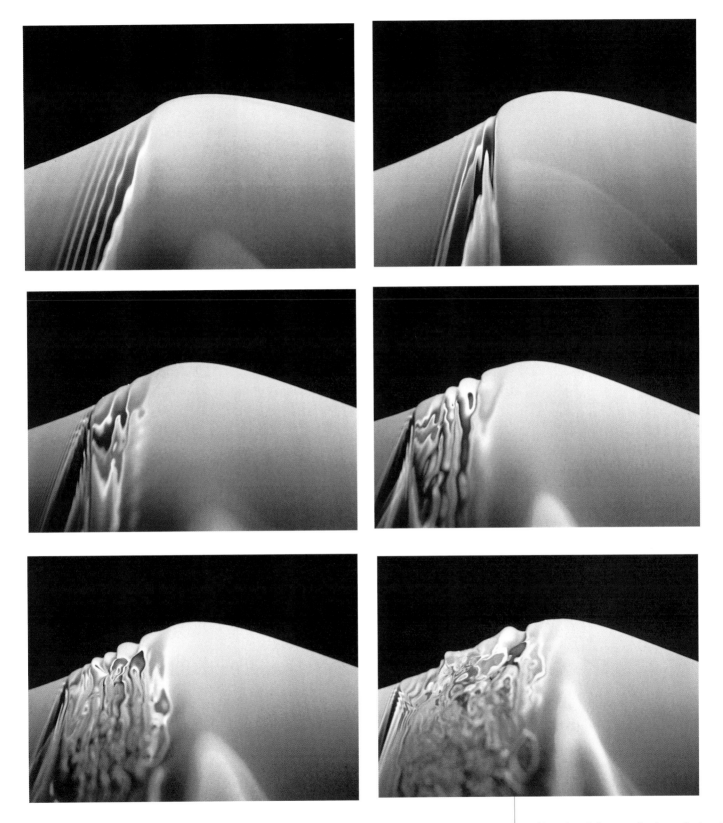

This series of photographs shows the transition from a laminar to a turbulent flow of a breaking water wave. The breaking process begins when a bulge forms on the forward face of the wave just below the crest. As the bulge moves down the face of the wave, a short wavelength disturbance appears and this disturbance soon breaks down into turbulent flow.

Yorke at the University of Maryland, and others. Mathematical physicists including Mitchell Feigenbaum at Los Alamos and Kenneth Wilson at Cornell added their ideas to the new amalgam of concepts.

Many theorists began to focus on the surprising mutation of systems as they change from equilibrium to chaos—a process that, at least superficially, appears to disobey the second law of thermodynamics. That law, as framed in the nineteenth century, decrees that order decreases over time. Yet many natural phenomena (for example, the generation of six-pointed snowflakes out of random arrays of water droplets in the air) seem to generate well-organized patterns as they progress toward chaos.

The onset of boiling is another familiar example. Put water in a saucepan until it is an inch deep and turn on the heat. Soon, peculiarly regular arrays of circular bubble patterns form as warm water rises, cools and sinks again. These "convection cells" have a large-scale meteorological counterpart in the remarkable structural similarity of thunderheads. Keep adding heat, and those symmetrical arrangements eventually blow themselves apart. But for a while, they seem to have created what Santa Fe Institute complexity theorist Stuart Kauffman calls "order for free"—a pattern that assembles itself.

There are many other analogues in nature. In chemistry, phenomena such as the Belousov-Zhabotinski reaction cause liquid reagents to alternate color as regularly as stop lights. Many theorists now suspect that a related form of chemical self-organization called autocatalysis—in which interacting compounds spontaneously create the conditions that sustain reactions—was crucial to the origin of life on Earth nearly 4 billion years ago.

The Natural Order of Things

Some researchers believe similar dynamics drive many oscillatory phenomena in nature, from the rhythm of heartbeats and brain waves to the periodicity of stars such as Cepheid variables, which vary regularly in brightness.

Interestingly, the same kinds of equations that are used to describe weather systems can be employed to calculate alternations in animal populations over time, which may undergo many "boom and bust" cycles before settling into equilibrium—if they ever do at all.

In medicine, researchers are studying heart attacks and seizures as chaotic perturbations. They are hoping that what has been learned about the subtle Butterfly Effects of initial conditions on chaotic systems can help them find ways to nudge a patient's heart rhythm or neural pattern back toward normal when it veers into chaos.

Thus, the "sciences of complexity" have united various aspects of physics, mathematics, chemistry, and biology in a common goal: understanding of how nature creates order—and exactly what kinds of order are permitted in the cosmos—on the smallest and largest scales.

A key field of inquiry is what happens at the boundary between two distinct states—the process known as a phase transition, perhaps most familiar in the change from liquid water to ice. Similar phenomena govern a number of events, such as the magnetization of iron below a certain temperature.

The Digital Laboratory

If the study of chaos and complexity owe much to mathematics and observation, they perhaps owe more to the new conceptual opportunities arising from computers. Before fast digital machines became available in the 1970s, the calculations necessary to visualize step-wise processes such as gradually evolving weather systems were laborious and time-consuming. It was extremely difficult to see how a complex system would change over long intervals, or to watch the interaction of dozens of variables.

But as powerful computers arrived in more labs, scientists were not only able to conduct broad-scale simulations requiring millions or billions of calculations, but were able to color-code the results and display them. Suddenly, emerging patterns were easily visible. By the 1980s, physicists were able to model such complicated processes as aerodynamic lift on an airplane wing. By the 1990s, they had begun to create reasonably accurate simulations of enormously intricate reactions such as nuclear fusion.

In some cases, such as the solutions to fractal equations, computer models were virtually the only way to visualize the results. In others, simulations were able to eliminate the need for entire suites of costly experiments. Thus if a simulation showed that a wing design was inadequate, there was no need to engage in labor-intensive wind-tunnel tests.

In other cases, detailed simulations were forced to take the place of experiments that simply cannot be conducted. The physics of stellar evolution and the potential emergence of global warming, for example, can only be studied in digital models. In addition, when the United States and other countries agreed in principle to a comprehensive nuclear test ban treaty, physicists were obliged to find alternative means to investigate the dynamics of fusion. Supercomputer models provided the means, although fusion experiments will continue with small quantities of fuel.

The physics of star formation and the mechanics of other astrophysical processes are now sufficiently well understood that they can be modeled in computer simulations and then refined by observations. This image shows a recently discovered phenomenon: as stars form, they often generate powerful, outflowing streams of interstellar gas. In this case, the process proceeds from left to right, and produces distinctive turbulence effects to the far right of the picture.

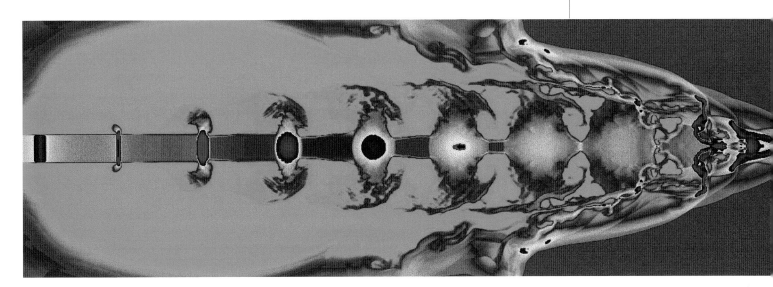

Surprising Shapes

Meanwhile, plenty of physical experiments were being conducted as different researchers were exploring the opposite of chaos: the way materials in extreme conditions make critical transitions to *more* orderly states.

It happens, for example, at almost unimaginable pressures when ordinary carbon in the form of graphite rearranges itself into diamond, a feat finally achieved artificially by General Electric scientists in 1954, at pressures of 100,000 atmospheres (about 1.5 million pounds per square inch) and a temperature around 1,600 C.

Using modern high-pressure apparatus made possible by the pioneering work of American physicist Percy Bridgman in the first half of the century, physicists are investigating what happens when hydrogen is squeezed as hard as it may be at the center of planets. At titanic pressures, theory predicts, hydrogen will achieve an orderly metallic state similar to its chemical cousins lithium, sodium, and potassium. Others are examining the way materials change their optical or electrical properties in enormous magnetic fields, or the precise threshold conditions that cause the onset of lasing.

Among the thousands of such research projects in the second half of the century, none was as astonishingly unexpected as the one that led—serendipitously—to the discovery of a new form of carbon called the buckminsterfullerene, or "buckyball" for short. In the process, it confirmed the recognition that mathematically complex, but highly orderly, states can arise out of even the most profoundly disorderly conditions.

Not that the researchers had set out to find a new molecular incarnation of carbon. In fact, no chemist or physicist thought there was any reason for carbon atoms to combine in symmetrical patterns other than the two familiar ways: graphite and diamond.

Instead, the research began with a question literally far removed from terrestrial considerations: namely, what process forms the long-chain strands of carbon and nitrogen that are seen in the outer atmospheres of some stars and floating in interstellar space? In the early 1980s, British researcher Harold Kroto had detected the spectral signature of those compounds and wanted to know how they might have been created. That, however, would mean finding a laboratory that somehow could replicate the superheated miasma around stars.

As it turned out, there was such a lab at Rice University, where chemical physicist Richard Smalley had built a laser-driven device that could vaporize almost any substance, turning the atoms into a seething plasma and permitting him to investigate the distribution and patterns of the atomic clusters that resulted when the plasma cooled and condensed. In 1985, Kroto came to Rice at the suggestion of Smalley's colleague and collaborator, Robert Curl. The three began to look at what resulted when carbon atoms were vaporized in a vacuum chamber and then settled out into clusters—a process presumably similar to what might occur in space.

By changing the conditions in the plasma chamber, they produced different kinds of atomic aggregates. That was to be expected: when a gas of atoms congeals, most of the atoms usually reform themselves into units of one or more particular arrangements

It is frequently difficult to observe the three-dimensional effects of regular or chaotic motion in laboratory samples. So physicists at the University of Texas at Austin took a novel approach to the problem. They induced vibrations of various sorts across a surface covered with tiny bronze spheres, each about as wide as a dime (2 mm). When a standing wave is generated, the spheres in low-energy regions, or nodes, scarcely move at all. But at regions of greatest intensity, the spheres are thrown into peaks like the one pictured here. Such standing waves are called "oscillons,"

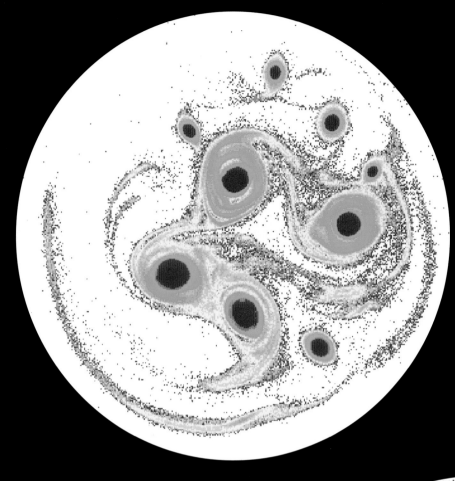

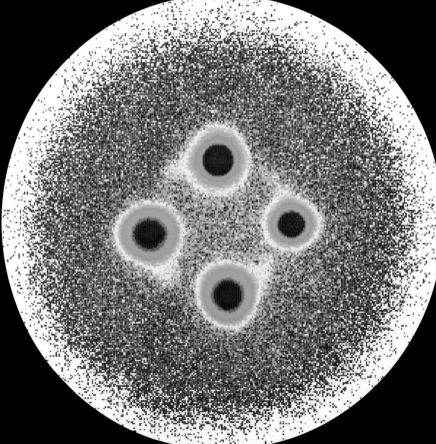

Whirlpool-like eddies, or vortices, arise in many well-stirred, or turbulent flows (above). Once the stirring stops, individual eddies tend to relax and merge into one large whirlpool. This picture shows a two-dimensional magnetized electron "fluid," which surprisingly does not relax into the expected single large whirlpool; rather, four vortices are formed in a geometric pattern and persist for a long time (right). The colors show rotation rate, with purple representing the highest spin rate and red the lowest.

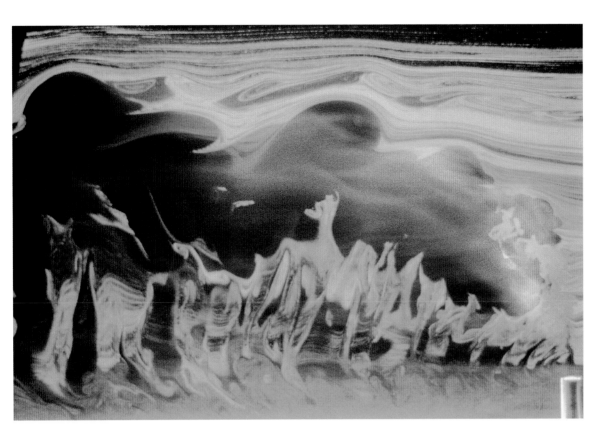

In this photograph, a stream of burning methane moves from left to right into a moving stream of air—a situation analogous to what happens behind jet engines on an airplane. As the flame front (blue, shading into yellow and red) moves to the right, whirling eddies of turbulence are formed in the air stream. Those whirling motions are indicated in green.

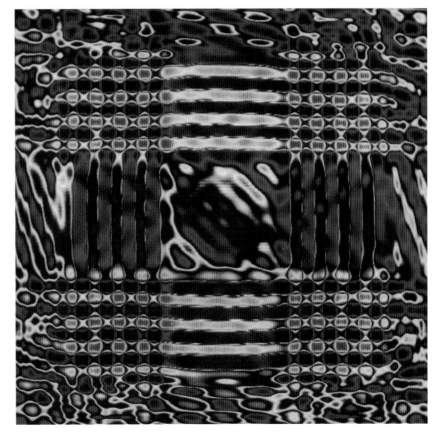

By using equipment that detects very small increases in the height of a surface, physicists can study the way waves interact. Here a regular pattern of waves results when a plate is struck in the center. A wave, originating from the center of the plate, reflects from the edges and soon waves are moving back and forth across the plate surface. A stable pattern is shown where waves produce the waffle-like interference array visible in this image, indicating places where the waves reinforce one another or cancel each other out.

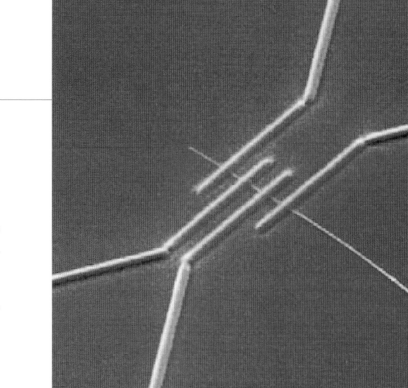

Just as natural processes can result in numerous forms of chaos, they can also sometimes create spontaneous order. A dramatic case in point is the way vaporized carbon atoms, under certain conditions, will assemble themselves into highly regular, cylindrical structures called nanotubes. Although carbon in its ordinary states is an insulator, physicists can manipulate nanotubes to alter their ability to conduct electricity. As a result, nanotubes might be employed as tiny circuit elements. Shown here is a single nanotube connected to four tungsten wires. Each of the tungsten wires is about the thickness of a human hair.

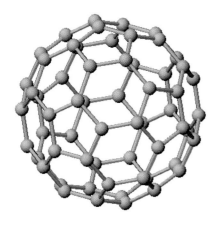

that are preferred by the laws governing atomic bonding. Often there are two preponderant cluster sizes—one small and one large—that reflect particularly symmetrical patterns of atoms called "magic numbers." These can be measured by using a mass spectrometer, in which a stream of individual atoms is propelled through a strong magnetic field that deflects the path of each one. The heavier the atom, the less its path curves in the field; so the final location of the atom in a detector is a very sensitive measure of its mass.

Smalley, Kroto, and Curl examined the soot from the plasma chamber, looking for the long-chain carbon compounds. What they saw was different and very odd. Under certain conditions, a large number of the carbon atoms had reassembled themselves into units of 60 or 70 atoms each.

What was going on? What configuration would produce a molecule of that size? One possibility was an icosahedron—a shape with 20 faces made up of alternating pentagons and hexagons that resembles a soccer ball or the geodesic domes favored by architectural designer R. Buckminster Fuller. The scientists didn't arrive at that arrangement by guesswork. It was well known that carbon atoms readily formed into

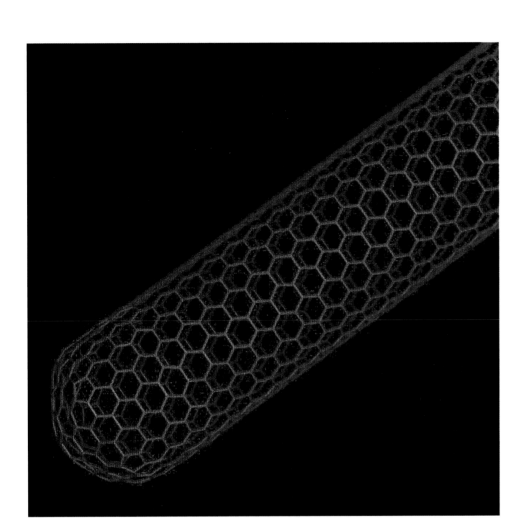

Shown here is a computer model of a carbon nanotube created by Richard Smalley of Rice University. Nanotubes, stiffer than steel, only a nanometer wide but many microns long, are essentially rolled-up sheets of carbon hexagons. They might become useful as microscope probes, as elements in integrated circuits, and as fibers in composite materials.

pentagons and hexagons, and the soccer-ball shape agreed with the geometrical construction principles for closed structures articulated by eighteenth-century Swiss mathematician Leonhard Euler.

If the molecules actually had that shape, they wouldn't be very reactive: each carbon atom would be bound to its neighbor by a single or double bond, leaving little opportunity for another element to attach itself. And, in fact, repeated tests showed that the newfound shape—dubbed the "buckminsterfullerene"—wasn't very reactive at all. Moreover, when carbon was vaporized in the presence of a metal vapor, the metal atoms became trapped inside the fullerene cage in a sort of "shrink-wrap" fashion. And the cages, which ranged in size from 44 to 60 carbon atoms, proved so durable that they were resistant even to strong pulses of laser light.

When the scientists published their findings in 1985, they met with considerable skepticism. No one had any reason to believe that carbon would form spherical or ovoid molecules with dozens of atoms. But by 1990 other researchers, using an electric arc between two carbon electrodes, managed to produce quantities of fullerenes so large that their existence became incontestable.

Soon labs around the world were experimenting with the unusual and promising properties of buckyballs. For example, large cages might be used as a biological delivery device for therapeutic chemicals. Equally exciting, fullerenes can be encouraged to form cylindrical "nanotubes" of extraordinary tensile strength. These eventually could be employed as sensitive tips for scanning tunneling microscopes, or combined in huge quantities to strengthen materials in much the same way that graphite fibers are now used. Furthermore, the addition of certain atoms makes nanotubes superconducting.

Going with the Flow

At the same time, a number of physicists were continuing to investigate another kind of spontaneous order that occurs when substances make a critical transition from one state to another. One of the most dramatic is the onset of superfluidity—a condition in which helium loses all internal friction, or viscosity. It not only produces some of the more visibly eerie effects in modern science, but shows that shifts from turbulence to orderly patterns are possible even at the quantum-mechanical level.

The study of superfluidity spans the entire century. It was first achieved by Dutch physicist Heike Kamerlingh Onnes, who succeeded in liquefying helium in 1908. Kamerlingh Onnes' lab then went farther, making the helium even colder than its liquefaction point of 4.2 K—that is, only 4 degrees Celsius above absolute zero. At the time, however, the researchers did not recognize the set of bizarre changes that take place in the liquid when it goes below 2.17 K.

Common sense suggests that liquids get more like solids as they get colder. But at 2.17 K, liquid helium undergoes yet another phase transition and suddenly flows without resistance. It can penetrate the tiniest apertures. If held in a cup, it forms a film that creeps up the inside surface, goes over the rim, and runs down the outside! At the same time, its density drops, its capacity to hold heat increases, and its ability to conduct heat skyrockets to millions of times the non-superfluid value.

The total effect of these changes was not known until 1938, when they were described by Soviet physicist Pyotr Kapitsa. But exactly why they happened remained a mystery. German-born physicist Fritz London and Soviet physicist Lev Landau provided initial theoretical explanations in the 1940s and 1950s. Landau's was the first to consider quantized states of motion for the whole fluid instead of individual atoms, as if the atoms were somehow working in concert, each in an identical condition. American physicist Richard Feynman later improved on the theory.

The context for those and all subsequent explanations derived from a prediction made by Albert Einstein in 1924. According to that suggestion, considered pretty outlandish at the time, under the right circumstances, all the constituents of certain substances can occupy the same quantum state with the lowest possible energy level. In other words, the wave function that defines the condition of every atom is identical, and in effect the entire system has the same wave function. It is in a highly ordered state.

That kind of behavior might appear to be impossible for matter particles such as protons, neutrons, and electrons—collectively known as fermions after Italian-born physicist Enrico Fermi.

A superfluid is a substance that, when cooled to a certain transition point, loses all resistance to flow. One way to think of it is that each atom has attained the same quantum state—the lowest energy state they can achieve—and so each cooperates with its neighbors. This spontaneous transition to a new order, which occurs abruptly about 2 K in liquid helium, produces some bizarre effects. For example, if a cup is dipped into a bath of superfluid helium and raised above the surface, the fluid will creep up the inside surface of the cup and drip down the outside. In this 1977 photograph, a container of superfluid liquid helium is heated very slightly, causing the contents to gush up in a perfectly symmetrical parabolic "fountain" effect.

A thin film of fluid flowing down an inclined plane does not necessarily do so in a smooth, orderly way. Sometimes the motion becomes unstable, causing peculiar "fingers" of fluid to form as the substance descends. Researchers captured this image by photographing a fluorescent fluid in ultraviolet light.

Fermions (which have half-integer spin values of 1/2, 3/2 and so forth) have to obey the Pauli exclusion principle described in Chapter 3. That is, no two particles in a system can occupy the same quantum state. The other class, which includes radiation-like particles such as photons, are under no such constraint. Those particles, called bosons, have integer spins, and any number of them can be in the same quantum state.

But sometimes, physicists Paul Ehrenfest and Robert Oppenheimer suggested in 1931, entities made up of even numbers of fermions can behave in the aggregate like bosons. Thus atoms of the most common isotope of helium (^{4}He), each of which contains two protons, two neutrons, and two electrons, could attain the same quantum condition. Because the spins of the atoms cancel each other out, giving the whole array a net spin of zero, they could condense into a common identical state, much like the Bose-Einstein condensation mentioned in Chapter 3.

By contrast, the isotope ^{3}He contains an odd number of particles, and thus should not—unless, that is, two ^{3}He atoms could somehow combine into coupled units, much like the "Cooper pairs" of electrons that make superconductivity possible. Theory suggested that it might be possible. But if so, it would entail reaching temperatures around 0.002 K, a thousand times colder than superfluid ^{4}He.

In the early 1970s, a team of three American physicists at Cornell, Robert Richardson, David Lee, and Douglas Osheroff, finally achieved that goal, thereby confirming Einstein's theoretical prediction, opening another connection between quantum mechanics and the visible world, and revealing one more way in which nature creates order at critical junctures.

Dozens of labs around the world are now using the same mathematical tools, derived from decades of chaos and complexity research, to study the spontaneous emergence of organization—or, alternatively, the collapse of order into chaos—in a stupendous variety of fields. These range from the sudden outbreak of epidemics, the extinction of species, and the distribution of wealth in populations to the literally largest question of all: how did the universe evolve into its present shape?

7

COSMOS

In 1900, the universe was still pretty much the way Newton left it. Traditional astronomy, studying the trickle of visible light from the heavens with increasingly powerful telescopes, had made enormous progress in observing and cataloguing celestial phenomena. And the ability to split that light into spectra, thus revealing the chemical content and temperature of its origin, had opened new and fascinating forms of analysis.

But there had been no truly profound innovations in cosmology since the seventeenth century, and even local conditions were considerably uncertain. As late as the mid-1920s, many leading scientists believed that our own Milky Way galaxy contained all the matter in the universe. Indeed, no one knew that there are nine planets in the solar system (Pluto would not be discovered until 1930), or even why the sun shines.

Nor could scientists plausibly answer the biggest questions of all. What *is* space and how big is it? What are stars and how do they shine? What elements can be found in the universe, and where do they come from? And when did it all begin?

Each of those ancient mysteries began to yield as astronomers applied the concepts and techniques of modern physics to conventional astronomical methods. The result would so drastically transform our ideas of the cosmos and Earth's place in it that by the end of the century it can fairly be said that humanity occupies a completely different universe from the one that was imagined only three generations earlier.

The Dawn of Relativity

The single most revolutionary conceptual change arrived in 1905, when Albert Einstein produced the theory of special relativity. As the name suggests, it applies to observations made in a particular special set of circumstances (those on the Earth's surface, for example) relative to any other moving systems such as planets or stars. Although it seems painfully counter intuitive at first, it abruptly eliminated one of the most worrisome problems of nineteenth-century physics: the constant speed of light.

Albert Einstein, shown here in 1907, made enormous contributions to half a dozen areas of physics. But he is best known to the general public as the creator of the theory of relativity. That revolutionary constellation of ideas about the shape of space and time, the behavior of light, and the nature of gravity transformed astronomy in the twentieth century.

Maxwell's equations and subsequent experiments had shown that light travels at only one speed irrespective of how an observer might be moving relative to the light source.

Yet this was not true of other kinds of wave motion, which obeyed the dependable additive principle known to Galileo. For example, a sound wave traveling at 1,000 feet per second toward an observer who is approaching at 500 feet per second has a speed of 1,500 feet per second relative to the observer. How could light behave differently?

Einstein's leap of insight explained how it could. He postulated that the speed of light is always the same to any observer in any reference frame. The price to be paid for this constancy is that space and time—the two factors that determine speed—turn out to be relative to the way the observer is moving. The conventional conception of simultaneity simply disappears, along with the reassuring Newtonian idea of a single absolute reference standard for space and time.

Thus, if your friend were to take off in an accelerating rocket, and you could somehow watch as she approached the speed of light, you would see two odd things happen. A clock moving with her would appear to slow down, and she and the objects around her would shrink in her direction of motion. (Also, her mass would approach infinity as she neared light speed—a speed that is unattainable for any object that has mass.)

Special relativity has by now become an integral part of physics. But it is general relativity—Einstein's grander and even more revolutionary theory—that allows us to attempt to understand the universe as a whole. One core idea of general relativity is that the presence of mass alters the nature of space and time around it. That is, the gravitational field of a massive object bends space in its vicinity, and time changes accordingly. In American physicist John Wheeler's memorable formulation, "matter tells space how to curve, and curved space tells matter how to move."

These were radical and bewildering notions and were by no means accepted at the time. But they did make certain very specific predictions. For example, light should bend by a certain small angle as it passes through the sun's powerful gravitational field. In 1919, during a total eclipse of the sun, British physicist Arthur Eddington measured the expected effect. What's more, general relativity explained a persistent anomaly in local astronomy: the orbit of Mercury around the sun varied by a small amount that was inexplicable in Newtonian terms, but it accorded perfectly with the idea of a planet traveling rapidly through the strong curvature of space close to the sun.

Einstein became an international sensation. (Nonetheless, he received the Nobel Prize in 1921 only for his work on the photoelectric effect; relativity was still too weird and insufficiently confirmed by experiment.) But there were ramifications to the theory that even its creator did not accept: lurking in the equations was the suggestion that the universe ought to be either expanding or contracting! Einstein dismissed this seemingly ridiculous effect by adding a term called the "cosmological constant." This was equivalent to postulating an outward pressure that countered gravity and kept the universe in a static, "steady-state" condition. Years later, he would call that act "the greatest blunder of my life." At the time, however, it seemed to make excellent sense.

Outward Bound

That was because, early in the twentieth century, astronomers had not yet broadened their horizons to the edge of the cosmos. And for good reason: there was no available "yardstick" to judge the extraordinary distances involved. For objects in our stellar neighborhood, an astronomer could use a parallax method, measuring the apparent position of the object at one point in the Earth's orbit around the sun, and then measuring again at the opposite point. The angle between the two readings would reveal the distance. But across spans of thousands of light years, that angle was immeasurably small.

Nor was apparent brightness much help. Astronomers had no way to know whether they were seeing an extremely luminous object that was very far away, or a dim object that was fairly close. As a result, it was perfectly plausible that the entire visible universe was contained in the Milky Way galaxy.

But by 1912, American astronomer Henrietta Leavitt had devised a system that would recalibrate the cosmos. The key was a certain type of star called a Cepheid variable that changed brightness on a regular schedule. Leavitt determined that there was a consistent relationship between the period of the star's pulsation and its brightness.

Thus when an astronomer located a Cepheid, its frequency of pulsation would indicate its absolute brightness. Its apparent brightness would show how far away it was because luminosity decreases as the square of the distance. A couple of years later, American astronomer Harlow Shapley used that method to demonstrate that the sun was not located at the center of the galaxy, as widely supposed, but was 28,000 light years from the center! That estimate proved to be within 20 percent of the currently accepted value. The universe was opening up.

Armed with the powerful Cepheid-variable technique, astronomers were soon to discover astonishing evidence of the very expansion that Einstein's cosmological constant had eliminated.

American Edwin Hubble had been using Leavitt's method, among others, to examine and catalogue various kinds of remote objects thought to be nebulae, clouds of glowing gas. He became interested in a phenomenon seen previously but never suitably explained: the spectral lines from some light sources appeared oddly skewed. They had the right characteristic features. But those features appeared in slightly wrong positions, as if the object's whole spectrum had been somehow shifted toward longer wavelengths. This "redshift" was intriguingly similar to the familiar Doppler effect, in which sound waves emanating from a rapidly receding object have a deeper pitch (lower frequency or longer wavelength) when heard by a listener who is stationary with respect to the object.

Hubble found that the farther away the objects were, the greater the redshift of their spectra. That is, the more remote the thing was, the faster it was moving away. (Clearly, the nebulae he was studying could not be part of the Milky Way. The apparent clouds were in fact other galaxies!) The specific numerical relationship between distance and recessional velocity, known as the Hubble constant, has been revised incessantly since it was formulated. The earliest version put the age of the universe at a mere 2 billion

One of the predictions of Einstein's general theory of relativity is that the wavelength of light changes as the light rays fall in a gravitational field. The Earth provides one such gravitational field, and so the effect (though incredibly small) should be measurable here. An experiment to test this prediction was devised by Harvard University physicist Robert V. Pound, and completed in 1960. Light was emitted from the top floor of the Jefferson Physical Laboratory and traveled 74 feet to the basement. The equipment shown above detected the light's wavelength at the top of the building. The apparatus shown right was installed at the bottom of the Jefferson Physical Laboratory, and used to measure the wavelength of the light after it had completed its 74-foot drop. Using extremely precise measurements, Pound and colleague G. A. Repka, Jr., were able to confirm Einstein's extraordinary prediction.

Violent events in the distant universe produce gravitational waves that eventually reach the Earth. The Laser Interferometer Gravitational Observatory (LIGO) is designed to detect such cosmic gravitational waves and harness them for scientific research. These waves carry information about their origins and the nature of gravity. A new window on the universe, LIGO is scheduled to begin operation in 2000. The interferometer shown here, in a laboratory at the California Institute of Technology, has L-shaped arms 40m long and is used for research and development required for the design of LIGO. When complete, the gravitational observatories, with arms 2.5 miles long, will be located in Louisiana and Washington.

When an alternate theory challenged the general theory of relativity, an experiment was devised to determine which theory was correct. To do this, the distance between the Earth and the moon was measured very accurately by shining a laser light beam at the lunar surface. The beam was reflected back from a mirror left on the moon by astronauts. The 1969 image above shows the laser light leaving the observatory. The experiment confirmed Einstein's general theory. The lunar ranging experiment has been repeated many times since 1969. In 1985, a new 30-inch telescope, opposite, was commissioned for ranging experiments.

years, younger than the Earth itself. Several researchers corrected Hubble's distance measures, and accuracy improved along with technology. But the fundamental relationship remains solid, and most astronomers have now reached a tentative consensus that the universe is about 13 billion years old.

Although Hubble was unaware of it at the time, his findings agreed with the work of two theorists—Russian Alexander Friedmann and Belgian Georges Lemaître—who independently had determined that Einstein's equations call for an expanding universe.

That concept, one of the most radically influential ideas since the notion of infinite space, was not readily accepted. In fact, it would take decades of results from astrophysics, particle physics, spectral analysis, and the fortuitous detection of an eerie universal radiation background before there was convincing physical evidence of this explosive idea.

Classes of Stars

Meanwhile, others were wrestling with the nature of stars. The nuclear fusion processes that cause stellar luminosity would not be known for a third of a century. But there was much to be learned by studying the surprising consistencies among observations. As early as 1905, Danish astronomer Ejnar Hertzsprung detected an extremely interesting relationship between the brightness of stars and their color: the whiter, the brighter. A decade later, American astronomer Henry Russell found the same relationship between luminosity and spectral type, as defined by the most prominent lines in a star's spectrum.

Cepheid variables are a class of star studied in 1908-12 by Henrietta Leavitt who worked at the Harvard College Observatory. As their name implies, Cepheids vary in brightness and the rate at which the variation occurs is directly related to their intrinsic brightness. Knowing the intrinsic brightness of a star, its distance from Earth can be calculated. This makes Cepheids the beacons of the universe. In the picture, a sequence of images shows the variation of a Cepheid located in the spiral galaxy M100. This particular Cepheid doubles in brightness over a period of 51.3 days. From this information, we conclude that the galaxy M100 is 51 million light years from Earth.

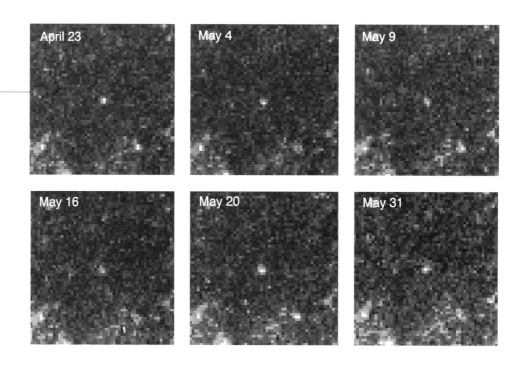

April 23   May 4   May 9   May 16   May 20   May 31

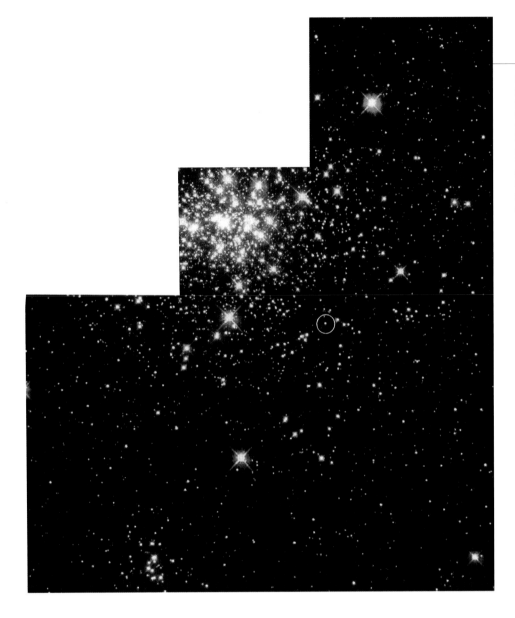

Astrophysics has been enormously successful in describing the processes whereby stars are born, age, and die. This photograph of a young star cluster some 164,000 light years away in the Large Magellanic Cloud reveals a stellar "nursery" containing about 20,000 stars. By studying the spectra of the light from these objects, physicists can determine their age and stage of growth.

It became clear that stars did not exist in an arbitrary variety, as many had thought, but only in certain readily discernible classes. And those classes were related in a uniformly predictable way when mapped in what were called Hertzsprung-Russell diagrams. Some were bright but red, indicating a relatively low temperature and, therefore, enormous size. Others were white but dim, suggesting that they were "white dwarfs" with surface temperatures in the tens of thousands of degrees. The vast majority made up a continuum between those extremes called the "main sequence," and their absolute brightness was directly correlated with their spectral type.

Our sun, a typical yellow star with a surface temperature around 5,800 C, turned out to be a particularly undistinguished specimen approximately in the middle of the range of stellar types.

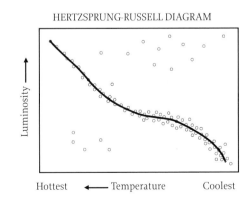

HERTZSPRUNG-RUSSELL DIAGRAM

Luminosity →

Hottest ← Temperature → Coolest

The Hertzsprung-Russell relationships implied that some constant physical principles governed the formation and evolution of stars. And they also provided yet another method of determining a star's distance from Earth: the distinctive spectral signature reveals the stellar type and luminosity class, which in turn corresponds to a particular absolute brightness. Comparing that to the apparent brightness yields the distance.

But why do stars shine at all? In the nineteenth century, two physicists—Lord Kelvin and Hermann von Helmholtz—had theorized that gravitational contraction of a star would cause its gases to heat up sufficiently to give off light. This explanation was plausible but problematic. For the sun to emit as much radiation as it does, it would have to be contracting so rapidly that only a few thousand years ago it would have had a diameter larger than the Earth's orbit. Worse yet, the Kelvin-Helmholtz model would give the sun a maximum lifetime of 10 or 20 million years, whereas geologists already had evidence that the Earth was about 100 times older.

Burning Issues

As mentioned in Chapter 5, in 1920, Eddington and colleague Robert Atkinson suggested a different process, in which hydrogen atoms combined into helium at the sun's core, which Eddington calculated must have a temperature of millions of degrees. Because a helium nucleus is slightly less massive than the sum of the separate masses of its protons and neutrons, the fusion process would convert the surplus matter to energy (in the form of radiation) in accordance with Einstein's $E = mc^2$.

But exactly how that might happen was unknown until 1938, when Hans Bethe worked out the details for both medium and large stars. Bethe postulated that in stars like the sun, with core temperatures below 16 million K, the chief mechanism was the direct combination of hydrogen nuclei into a heavy hydrogen isotope called deuterium and then into helium and, eventually, a few other light elements.

More massive stars could produce much heavier elements, as spectral analyses showed. Bethe accounted for this phenomenon by postulating a carbon-nitrogen-oxygen (CNO) process: in stars with core temperatures above 16 million K, carbon acts as a catalyst, absorbing hydrogen nuclei and transmuting into nitrogen and then oxygen before throwing off a helium nucleus and returning to carbon again to continue the cycle.

The CNO process could easily explain the creation of elements up to the mass of iron. (As noted earlier, iron is the most stable of all nuclei. Thus, any process that makes heavier elements from iron must draw on some other source of energy rather than producing it from the change itself.) As the century progressed and particle physics came to play an increasingly crucial role in astronomy, it became clear that heavier elements are created in supernovas: the sudden death-collapse and subsequent explosive rebound of stars that have suffered fusion burnout. In those cases, the outward pressure of radiation no longer counterbalances the star's gravitational contraction. It implodes to the point at which it cannot be compressed farther and then blows up. The energy released forms heavy elements from the star's components and emits a flash of visible light and a torrent of neutrinos that can be observed from Earth.

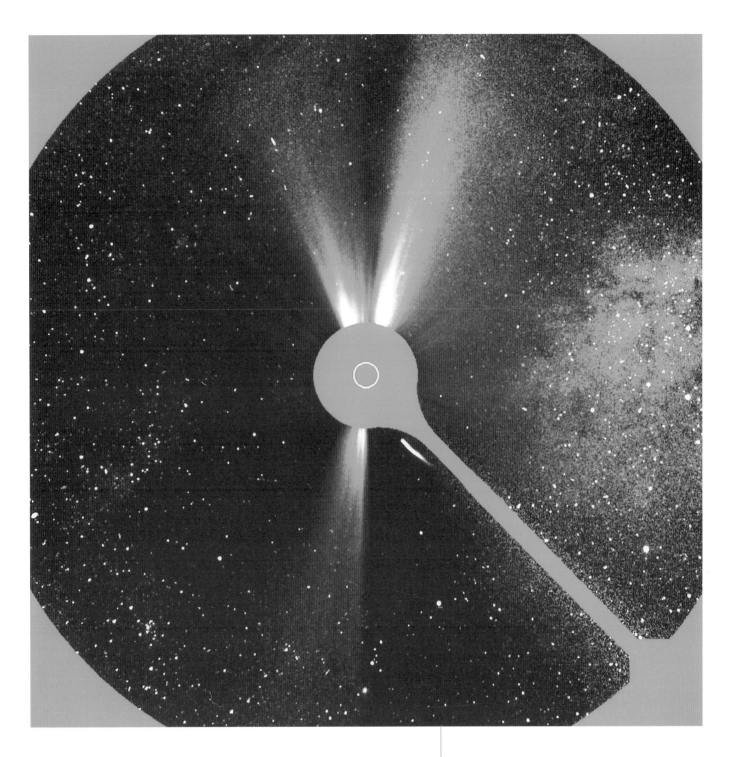

Throughout the twentieth century, increasingly sophisticated observations and theories have made the physics of the sun reasonably well known. But there are still peculiar phenomena that are poorly understood. One of them is the explosive emission of mass from the sun's outer atmosphere, or "corona." In this 1996 image, taken by the orbiting SOHO (Solar and Heliospheric Observatory) satellite, the sun's circumference is blocked by an "occulting" disk in order to make the "coronal mass ejection" more clearly visible. The ejections above and below the disk extend thousands of miles into space, presumably driven by fluctuations of the magnetic fields in the solar surface. The curving line to the bottom right of the sun is a comet.

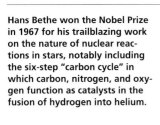

Hans Bethe won the Nobel Prize in 1967 for his trailblazing work on the nature of nuclear reactions in stars, notably including the six-step "carbon cycle" in which carbon, nitrogen, and oxygen function as catalysts in the fusion of hydrogen into helium.

Bethe's notion not only explained the spectra of stars, but permitted stellar ages lasting billions of years. Others added to and refined the concept. Indian-born Subrahmanyan Chandrasekhar determined the stages of stellar evolution, incorporating Einstein's relativistic effects. British physicist Fred Hoyle extended Bethe's CNO theory to show that it requires carbon-12 to assume a certain energy level; in 1957, William Fowler and colleagues at Caltech found precisely that level in the lab—confirming Bethe's conjecture and solidifying the concepts of stellar dynamics that are still in use today.

A Recipe for Creation

With Hubble's redshift observations and a solid theory of star physics, the modern cosmos was beginning to take shape. By midcentury, however, it was still unclear whether the universe might be changing over time, and how that idea could account for the observed abundance of elements.

There were two main theories. One, championed by Hoyle, posited a "steady state" in which the Hubble expansion occurred, but some unrevealed process converted energy to matter in the new space in just the right way to leave the average density of the cosmos unchanged over time. In the other model, which Hoyle derisively called the "Big Bang" theory (a name that stuck), the universe had exploded outward from an initial hot, high-density state, cooling and expanding as it went.

It might seem impossible to find empirical methods to choose between such sweepingly different visions. But the new discoveries of nuclear processes provided stringent criteria for the ways and circumstances in which particles can combine. Cosmology had become inseparable from particle physics. Hoyle attempted to accommodate the criteria by positing that all nuclei are formed in stars. But Russian-born theorist George Gamow produced a powerful alternative explanation. He determined in the 1940s and 1950s that an early Big Bang cosmos would have had temperatures and densities high enough to cause particles to clump together into certain limited kinds of nuclei—a process called nucleosynthesis.

In concert with American colleagues Ralph Alpher and George Hermann, Gamow calculated that the laws of particle combination would have allowed the Big Bang to produce hydrogen and helium, together with a vanishingly small sprinkle of lithium, but no other heavier elements. The proportions predicted by the theorists turned out to agree closely with observed percentages, to the detriment of the steady-state theory. About one-fourth of the matter in the cosmos is helium-4. If all visible matter had been made in stars, no process known to physics could account for that much helium. Moreover, the universe contains a small but measurable quantity of deuterium, a heavy isotope of hydrogen. But stars cannot make anywhere near the observed abundance of deuterium because they burn it very readily. Indeed, deuterium would be almost absent in a steady-state cosmos.

Things were beginning to look better and better for the Big Bang. But a final, and crucial, discovery would tip the balance. It came by way of another of Gamow's predictions, one that was not regarded as critically important at the time: The earliest formation of atoms would have left a very distinctive photon signature on the heavens. That would happen because, in the emerging picture of cosmic origins, the universe evolved in stages. It began as a seething stew of elementary particles and radiation at stupendous temperatures, and cooled as it expanded. Finally, about 300,000 years after the expansion began, when the average temperature was at white heat and the universe was perhaps 1/1,000 its present size, electrons were finally able to start binding to protons, forming ordinary atoms of hydrogen.

Space suddenly became transparent. Until that time, the density of the separate electrically charged particles was so high that photons scattered at all angles as radiation excited the components of the universal plasma, which in turn emitted its own radiation. In effect, this made the cosmos opaque. But when the temperature dropped and atoms began to form, the photons which had been madly ricocheting throughout the particle miasma were free to travel unimpeded. Even billions of years later, Gamow suggested, those relic photons should be observable. They would have been drastically redshifted by the expansion of space, which would have stretched out their wavelengths. But they should still be around somewhere.

Echoes of the Bang

They were found quite by accident, and with that discovery the entire field of radio astronomy—the technique of observing emissions in other wavelengths than visible light—began. In 1933, Karl Jansky of Bell Telephone Laboratories in New Jersey was using a large antenna to investigate sources of troublesome static and interference in radio communications. To his surprise, he found that a lot of noise appeared at about the same time every day. It turned out to be perfectly in tune with sidereal time (determined by the position of the far away stars), in which the day is 4 minutes shorter than the solar day (determined by the position of the sun). He was, in fact, picking up radio waves from the center of the galaxy!

Thereafter, astronomers began listening to the heavens at a variety of wavelengths. In 1964, a pair of Bell Labs researchers, Arno Penzias and Robert Wilson, were trying to eliminate a source of background static from their radio telescope. The problem would not go away. Oddly, it was the same microwave frequency at the same strength in every direction, and it remained unchanged at all times of day or night, winter or summer. So if it was coming from space, it was originating well beyond our galaxy.

After consulting with several astrophysicists, they finally concluded that their microwave signal was Gamow's long-awaited cosmic background. It would take thirty-five years before a NASA satellite called the Cosmic Background Explorer (COBE) confirmed that the radiation was in fact isotropic (uniform in all directions) to within a fraction of 1 percent.

But isotropy, as it turned out, raised other troubling problems. Basically, the distribution of matter in space was both too homogenous, and too lumpy, to agree comfortably with theory.

The classic Big Bang model would have produced a rapidly expanding cosmos, but a nonuniform one—a cosmos far more clumped and cluttered than the smoothness revealed by the microwaves. The 1990 COBE measurements had shown that in all directions, the universe is filled with microwaves of exactly the same energy, as if they had been given off by an object with a temperature of 2.73 K. That, in turn, suggests that all the various parts of the cosmos must have been relatively close to each other at one time in order to have reached such a universal equilibrium temperature.

Yet looking out in opposite directions in the sky, we observe regions of space that are so far separated that even light could never have traveled between them during the life of the universe.

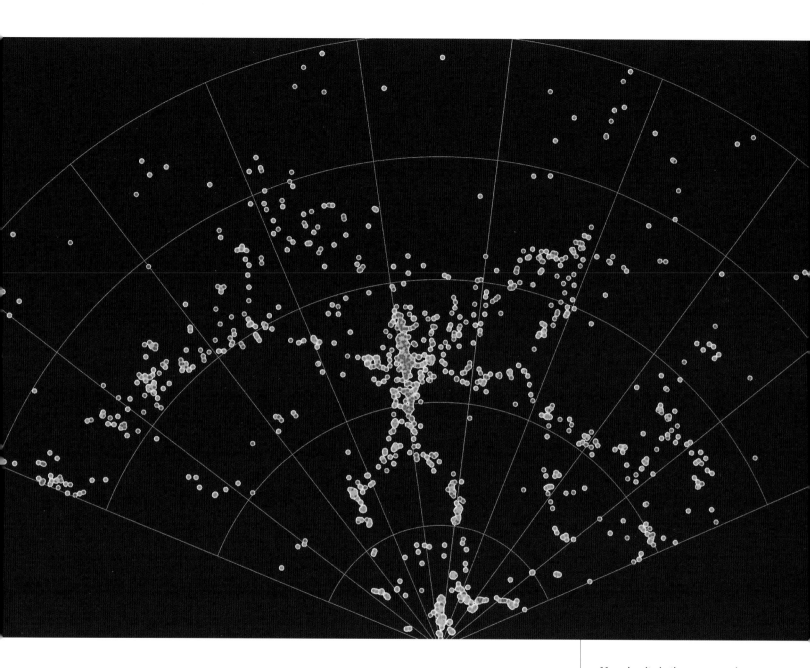

Mass density in the cosmos varies locally (as galaxies), regionally (as clusters of galaxies), and on even greater scales as clusters of clusters. By mapping the mass distribution, scientists hope to get a broader sense of how and why the universe has the structure it does. This image shows a three-dimensional map made by Margaret Geller, John Huchra, and colleagues at the Harvard-Smithsonian Center for Astrophysics, revealing the distribution of galaxies of various brightness values at a distance out to about a billion light years from Earth. If you look carefully, there appears to be a stick figure of a person in the center of the image.

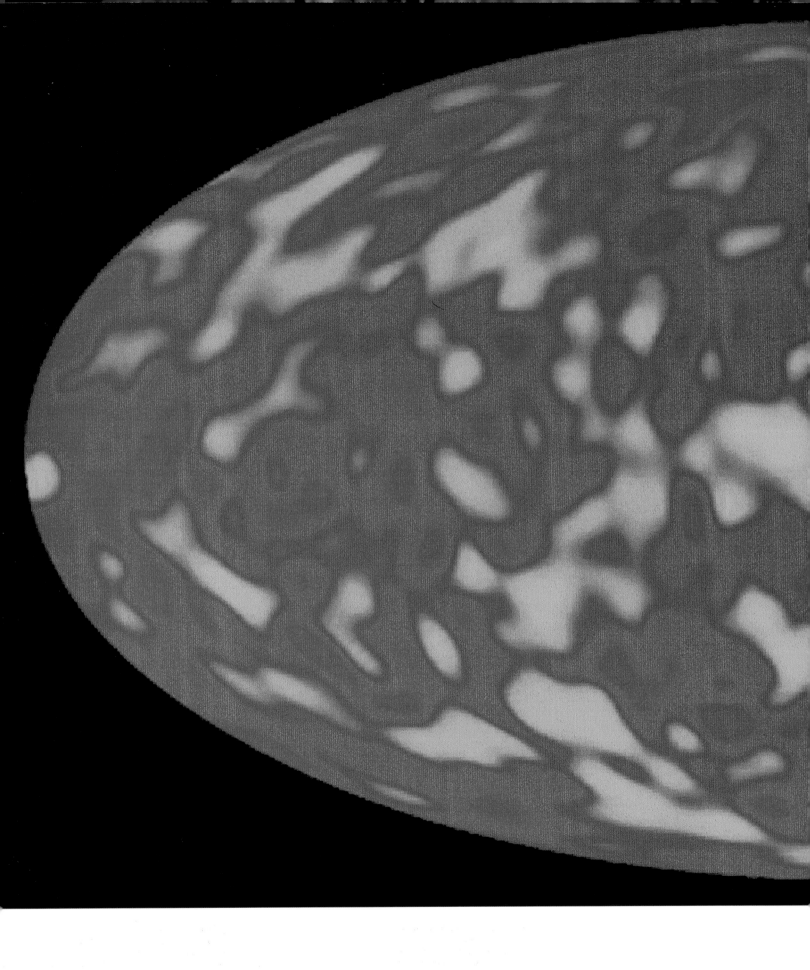

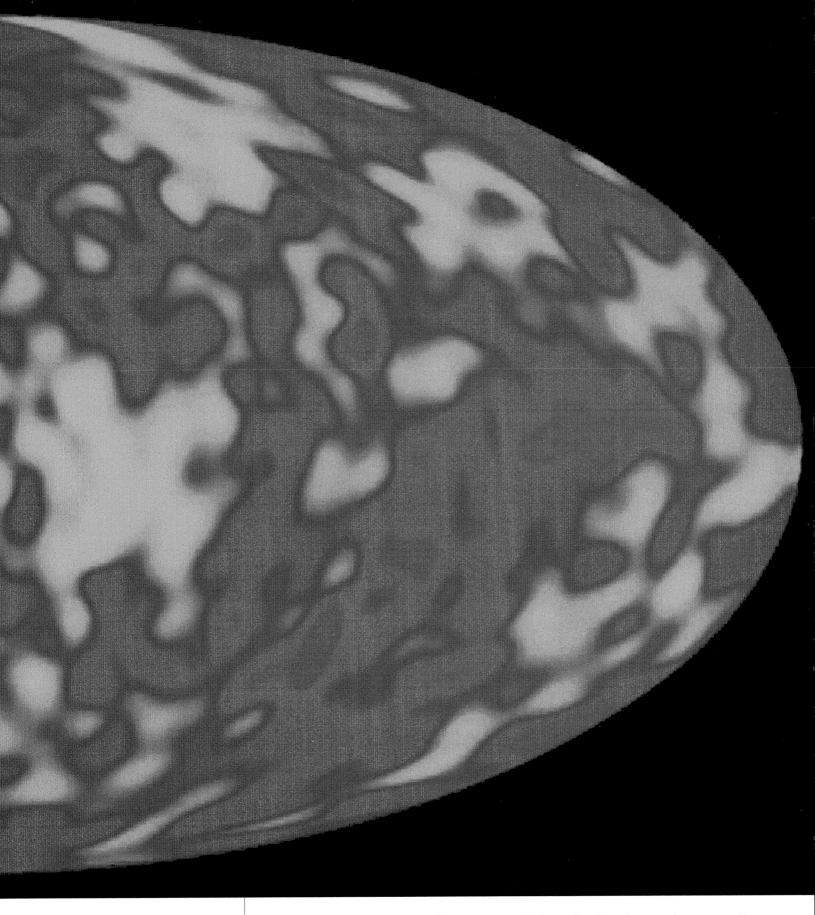

Most physicists now believe that our universe was created about 13 billion years ago in an event that has come to be called the Big Bang. There are several lines of evidence supporting the Big Bang hypothesis. One of the most compelling is the presence of an almost perfectly uniform background radiation everywhere in the sky, suggesting that the universe has expanded more or less homogeneously from a single point. This "cosmic background," a faint microwave glow corresponding to a temperature of about 2.73 K, was first detected in 1965. In 1992, highly precise satellite measurements by the COBE Science Working Group at NASA's Goddard Space Flight Center found that the background radiation varied by slight amounts in different sections of the universe, corresponding to irregularities in density in the early universe. This image shows those variations.

Similarly puzzling is the fact that the general distribution of mass is virtually identical in every direction. Moreover, the expansion seems to be more or less that of a "flat" universe—a term signifying that the expansion will never stop and reverse itself, but will continue at a constant rate.

What conceivable process could explain those two outcomes? In 1980, American physicist Alan Guth devised a solution, known as "inflation," that combined quantum mechanics, particle physics, and field theory. In Guth's scheme, which has been improved continuously since then, at a time about 10^{-35} seconds after the Big Bang, conditions were such that the cosmos began to double its rate of expansion every trillionth of a trillionth of a trillionth of a second. This process lasted only until 10^{-33} seconds after the bang, but drastically extended the universe while leaving its distribution of matter and energy nearly unchanged.

Inflation took care of homogeneity and "flatness." But as a quick look into the night sky will show, matter is not arranged with perfect uniformity in the heavens. There are not only clusters of stars, but galaxies and clusters of galaxies. There are even clusters of clusters. One of the most famously uneven sections of the cosmos is the "Great Wall"—an area in which some 1,100 galaxies form a sort of sheet about three-fourths of a billion light years wide—discovered by Margaret Geller, John Huchra, and colleagues at the Harvard-Smithsonian Center for Astrophysics.

So although the microwave background—which reflects the gross structure of the cosmos—should be almost perfectly homogenous, it should also vary from place to place, reflecting the original irregularities that existed at the end of inflation, which resulted in the local clumping of matter. In the mid-1990s, a refitted and improved COBE satellite revealed exactly that structure, to the considerable satisfaction of theorists.

Oddballs of the Cosmos

Microwaves, however, were by no means the only frequencies of interest to astrophysicists. As radio astronomy improved dramatically in the last third of the century—aided immeasurably by British physicist Martin Ryle's development of the "synthetic aperture" technique in which many small adjustable antennas can achieve the precision of a single antenna miles in diameter—observations uncovered a number of celestial entities that could only be explained by recourse to theoretical physics.

The Hubble Space Telescope, launched into orbit three hundred miles above the Earth in 1990, is unburdened by the planet's atmosphere, and has produced images of unprecedented clarity. Planned improvements and NASA's next-generation telescope should provide far greater precision in observing distant objects. This photograph shows the Hubble Space Telescope in the final stages of construction.

One was the quasar, or quasi-stellar object, which gives off radiation equivalent to a trillion suns from extreme distances. The most reasonable explanation so far is that they result from supermassive "black holes" within forming galaxies. Another was the existence of pulsars, strange and unexpected objects that emitted radio waves at precise intervals of about 1 second. First observed by British investigators Jocelyn Bell and Anthony Hewish in 1967, pulsars were eventually determined to be neutron stars, stellar cores that had condensed so drastically that electrons and protons collapsed into neutrons. A typical neutron star, of the sort postulated by American physicist Robert Oppenheimer and others in the 1930s, contains two times the mass of our sun in a sphere of about 15 miles in diameter. A teaspoon full of those compacted neutrons

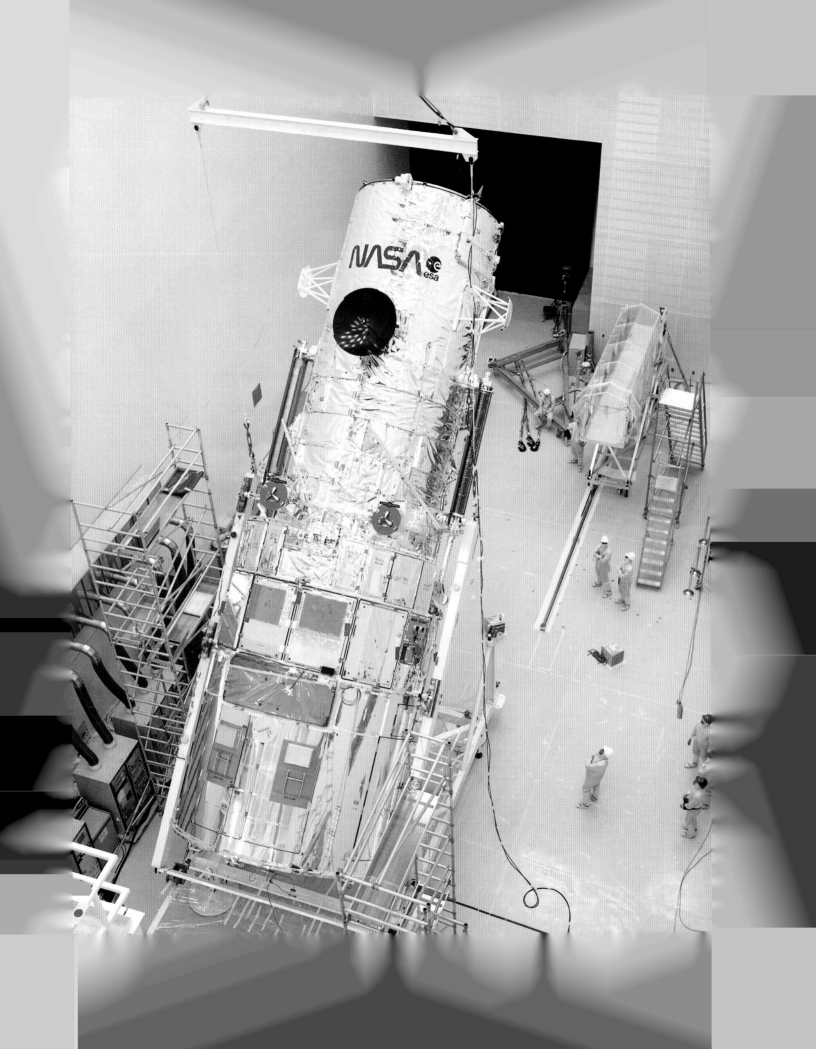

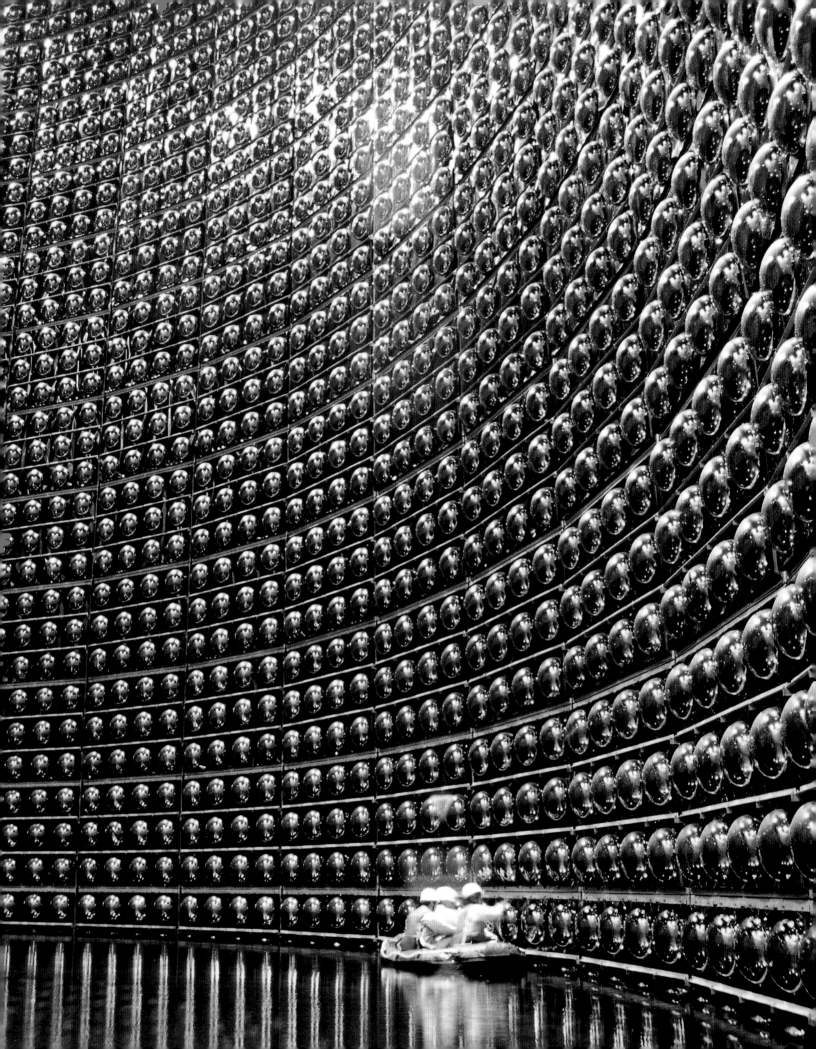

would weigh millions of tons! Presumably, the neutron star's rotation produces a powerful magnetic field that channels radiation in a beam that sweeps across the sky somewhat like the beacon of a lighthouse.

Twenty years later, astrophysicists expanded their arsenal of observational tools beyond electromagnetic radiation to include the elusive particles called neutrinos. By 1987, several detectors around the world were monitoring neutrino activity. One group of researchers was attempting to validate the accepted model of solar physics by measuring the neutrino output of the sun, which appears smaller than theory predicts. Other detectors were being used to look for evidence that protons could decay spontaneously. Some theories predicted that the stable life of a proton would be around 10^{32} years. Therefore, one proton out of every 10^{32} ought to decay every year on average.

In 1987, two big detectors—one in Japan, the other in a salt mine beneath Lake Erie—showed a sudden fury of activity far beyond expected levels. That burst coincided with the discovery by Canadian astronomer Ian Shelton of a huge supernova in the Large Magellanic Cloud. By remarkable good luck, it was the brightest supernova seen for centuries. The neutrino volley had arrived about three hours in advance of the visible light from the event, precisely as theory predicted: the neutrinos should be generated by nuclear processes in the collapsing stellar core that precede release of light. Neutrino astronomy became a serious research field. Physics had provided yet another way to "see" into the heavens.

Missing Mass and Black Holes

But perhaps no astronomical puzzle taxed the wits of physicists as much as what *couldn't* be seen. As early as the 1930s, astrophysicist Fritz Zwicky had discovered that there must be more stuff in the cosmos than we can observe at any electromagnetic frequency.

The evidence was powerful. The rotational speed of different parts of galaxies is strictly determined by Newtonian laws: the farther from the center of rotation, the slower the galactic components should be moving. Yet in our own galaxy, and in others later observed, that is plainly not the case. The outermost visible stars and clusters are moving faster than they should compared to those near the center—precisely what would be expected if the galaxy were actually much larger and at least ten times more massive.

Subsequent studies of star clusters and even clusters of galaxies confirmed the phenomenon. The components of those aggregations were moving so fast that they should have whirled themselves apart long ago. Yet they remain bound together gravitationally because they are apparently surrounded by a huge invisible halo of . . . something. Eventually, it became clear that at least 90 percent of the matter in the universe is "missing," in the sense that we have no way to detect it.

Physicists are still struggling to come up with candidates for this "dark matter," including superdim brown dwarf stars, massive neutrinos—a possibility reported in 1998 from a giant detector in Japan—and a roster of exotic hypothetical particles such as WIMPS (weakly interacting massive particles). Dark matter also has profound cosmo-

Conventional telescopes detect visible light. Radio telescopes detect radiation at longer wavelengths. But photons are not the only physical entities streaming to Earth from outer space. The planet is constantly bombarded by the elusive, near-massless particles called neutrinos. They interact so weakly and so infrequently with ordinary matter that they cannot be observed directly, but only by their collision effects on atoms. To detect neutrinos, physicists have built enormous underground chambers (to filter out other cosmic radiation), filled them with water, and lined the walls with sensors that pick up the distinctive light bursts that occur on the rare occasion that a neutrino strikes an atom. Shown here is the giant Super Kamiokande neutrino detector, buried one kilometer under a mountain near Tokyo, Japan, that recently produced evidence for neutrino mass. The Super Kamiokande neutrino facility contains 50,000 tons of highly purified water and 13,000 photomultiplier units to detect the telltale light signature (the optical equivalent of a sonic boom for sound waves) resulting from a neutrino collision. As the detector was filling in 1996, the technicians shown here moved around by raft to clean the faces of the photomultipliers before they were submerged.

Stars, planets, and sundry interstellar objects have their origin in volumes of gas and dust called nebulae (from the Latin for "cloud"). Shown here is the so-called Horsehead Nebula, an extremely dense cloud in the constellation Orion. The pinkish glow behind the nebula is emitted from ionized gas.

Towers of seemingly sculpted gas in the Eagle Nebula serve as cocoons for young stars. This image was taken by the Hubble Space Telescope in 1995.

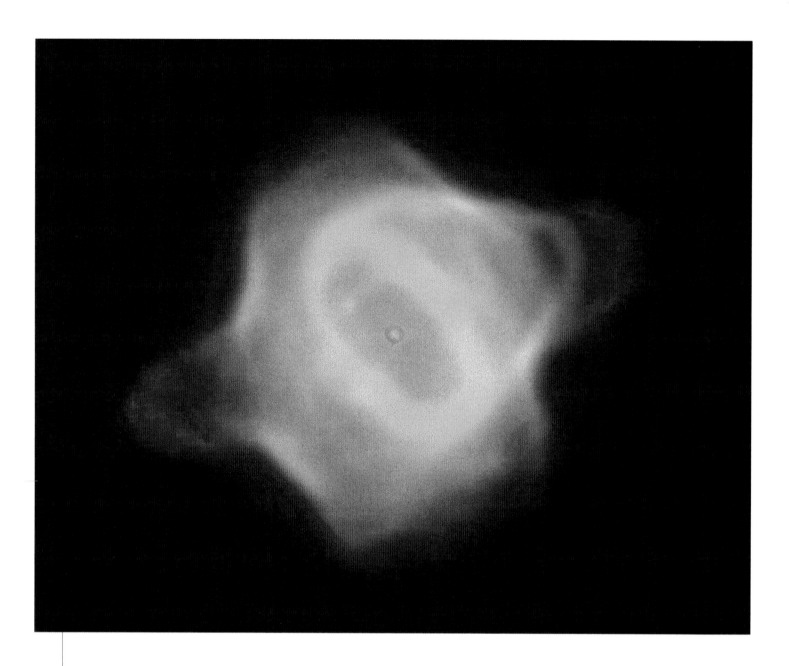

This 1998 Hubble Space Telescope image shows the Stingray Nebula, the youngest known of a class of celestial objects called planetary nebulae. That term is a bit misleading. The cloudlike material surrounding the central star does not form into planets. Rather, the clouds of dust and gas were ejected from the outer envelope of the star—in this case, within the past forty years. Eventually, the Stingray's central star will evolve into a white dwarf.

logical purport, because the amount of it in whatever form it takes will determine the total mass density of the universe, and thus decide the question of whether the cosmos will continue expanding indefinitely, remain about the way it is, or slow down and ultimately collapse back upon itself.

Although describing and detecting dark matter are formidable problems, twentieth-century physicists have learned to cope with many phenomena at the edge of the unthinkable. The most outrageous example is the black hole, an entity predicted by Einstein's theory of general relativity and first described theoretically in 1916 by German astrophysicist Karl Schwarzschild. Black holes seem to occur when stellar collapse is so powerful that matter is compacted, in theory, to infinite density at a single point. The resulting gravitational field is so strong that even photons cannot evade the pull. The fact that light cannot escape is what makes black holes black. Consequently,

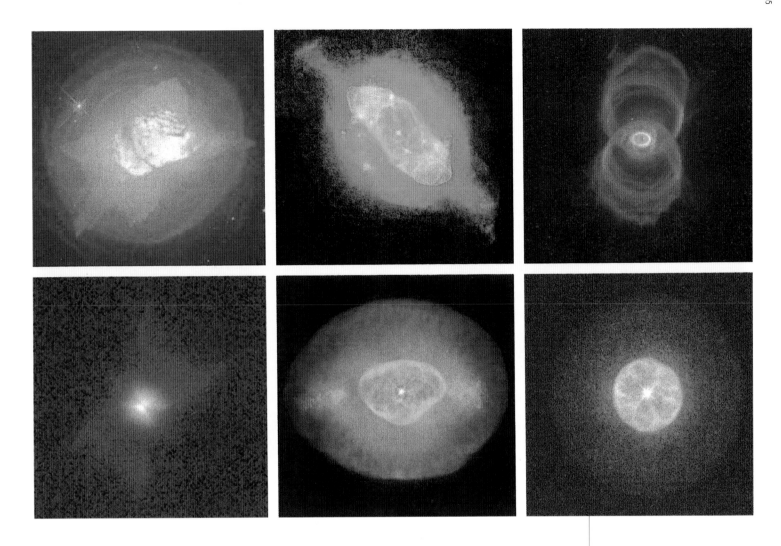

Planetary nebulae are typically beautiful structures. These six pictures of planetary nebulae were taken by the Hubble Space Telescope.

black holes cannot be observed directly, though Chandrasekhar set limits on the stellar mass needed to form one and New Zealand mathematician Roy Kerr described how they must rotate.

But black holes—thought to lurk at the centers of many radiation-emitting galactic cores, including ours—can be detected by their effect on surrounding material. As matter is accelerated en route to the all-devouring hole, it flows inward and is heated to the point of giving off strong X-rays and other high-energy radiation. Quite a few likely candidates are being found in collapsed stars and galaxy cores.

Thus in 100 years, the grand cycle of mystery, comprehension, and the discovery of further mystery has come full circle. Just as the twentieth century had begun with the search for the unknown constituents of the atom, the twenty-first arrives with physicists hunting for the fundamental components of the cosmos. If the past is a reliable guide, they will succeed. Nature's store of secrets is vast. But so is the indomitable curiosity of the human mind.

Galaxies are observed to have several distinctive shapes. On this page is a barred spiral galaxy (NGC 1365) and, opposite, a spiral galaxy (Messier 83).

Supernovas are cataclysmic events that occur in the late stages of a large star's life. In this event, the outer shell of the star is blown away from the stellar core. The most recent nearby supernova, 1987A (this page), occurred in the Large Magellanic Cloud visible from the southern hemisphere. A supernova in Vela (opposite) occurred about 10,000 years ago so the ejected material has had time to disperse more widely.

The Hubble Space Telescope shows a region near a dying star. Astronomers have dubbed the tadpole-like objects in the upper right-hand corner "cometary knots" because their glowing heads and gossamer tails resemble comets. Although astronomers have seen gaseous knots through ground-based telescopes, they have never seen so many in a single nebula.

The constellation Orion is one of the favorite objects in the winter sky. In that constellation is the Orion Nebula. This photograph shows the Orion Nebula taken with three frequencies of light.

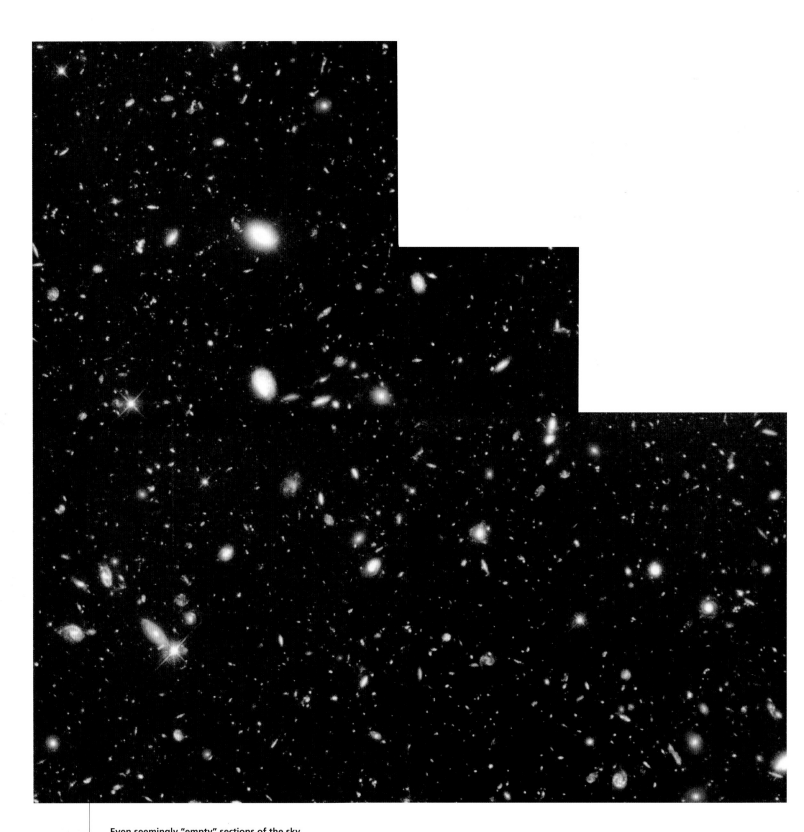

Even seemingly "empty" sections of the sky
are revealed to be full of galaxies when
viewed at extreme sensitivity. This picture
was made by aiming the Hubble Space Telescope at an apparently dark "deep field" near
the handle of the Big Dipper. It shows some
1,500 galaxies in various stages of development, including a few that may be colliding.
The area enlarged here would appear to the
naked eye about equal to a grain of sand held
at arm's length.

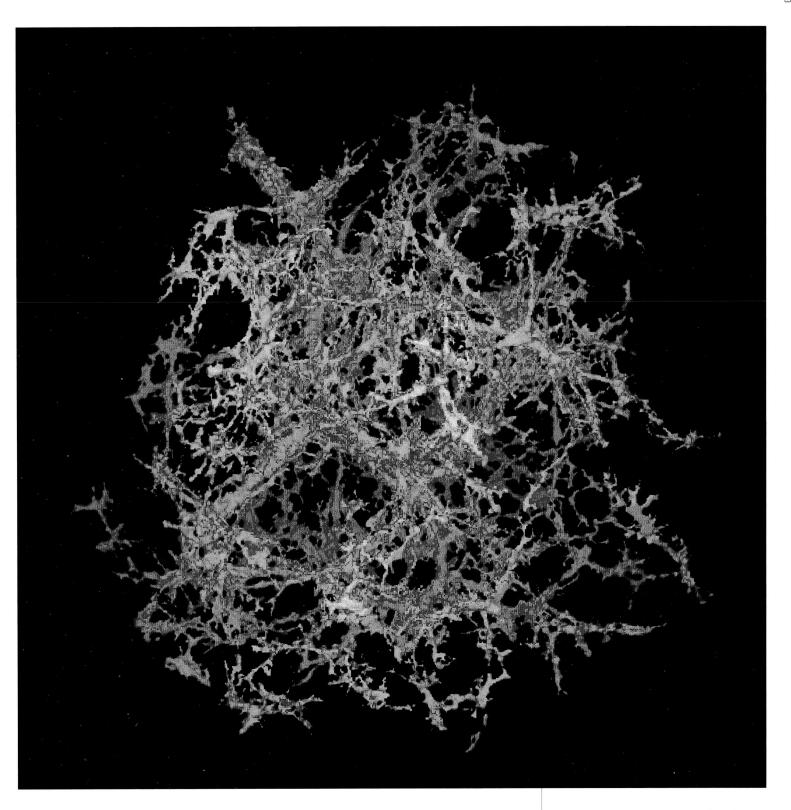

Some theories—such as the way the universe took shape after the Big Bang—might seem untestable by definition. But now supercomputers allow scientists to conduct "experiments" in cosmic formation. This image was produced by a program that uses the laws of physics to simulate the distribution of matter in the universe as it evolves dynamically over time. It shows the gross structure of a representative part of the universe at its present age.

CONCLUSION

It has been, by any standard, a century of incessant marvels. So it might be difficult for the bedazzled citizen to imagine that, over the next ten decades, physical science could possibly equal the revolutionary impact of the work done in the preceding ten. History, however, suggests that this may be the case: progress begets more progress. Isaac Newton once wrote that if he had seen farther than others, it was "because I stood on the shoulders of giants." Researchers of the twenty-first century will scale even greater heights because those of the twentieth provided an unprecedented perspective. Science is not some set of specific goals to be achieved once and for all; it is a never-ending, ever-improving, self-correcting means of expanding human understanding of nature.

Physicists are looking forward to an era of accelerating insight as investigative technology grows more powerful, experiments more sophisticated, and theory more profound. They will have to, because many fundamental mysteries persist. Even such seemingly familiar components of the Newtonian cosmos as mass and gravity have proven enigmatic as each new discovery has opened up even deeper puzzles.

For example, physics has traced the basis of all matter to two sets of apparently irreducible elementary particles called leptons and quarks—objects so small (no larger than 10^{-18} meters) that they are in effect dimensionless points. Yet the measured mass of the electron is less than that of the same-sized top quark by a factor of a million! And the three kinds of neutrinos, which make up half of the leptons, are at least 10,000 times lighter than the electron. Until 1998, most theories assumed that they had no mass at all.

This strange arrangement suggests that there is some mechanism, perhaps analogous to electric or magnetic fields, that confers the property of mass unequally among various particles. If so, that mechanism must have an associated particle that mediates its effect just as the photon conveys the electromagnetic force. The search for this particle—presumed to be something called the Higgs boson, after Peter Higgs, the Scottish theorist who conceived of it—will be one of the great challenges for physics in the coming century.

So will the hunt for another "missing" boson: the graviton that is thought to convey the gravitational force. Numerous very large devices to determine its properties are under construction. But even if the graviton is observed and characterized, a more formidable task will remain—to integrate gravity with quantum mechanics. If that is possible, it will give humanity a single consolidated view of the forces that shape the cosmos from the tiniest subatomic scale to the intergalactic.

Even that, however, will not be a complete view. Physics seeks to describe the most essential processes and relationships in nature. And if there is one overriding lesson that physicists have learned in the past two hundred years, it is that there is usually an underlying simplicity beneath apparent and often bewildering diversity. Thus, in the nineteenth century, investigators determined that electricity and magnetism were complementary aspects of the same phenomenon; and in the twentieth, they found that electromagnetism was a manifestation of the same process that produces the weak force governing radioactive decay.

Those and similar revelations lead naturally to the suspicion that there may be a single principle that governs and generates all the forces—and all the different kinds of particles—as it acts in different environments and circumstances. Such a principle may have been manifest in the early moments of the Big Bang, but may now be visible only in secondary forms, including the four forces and twelve particles that it devolved into as the universe cooled and congealed into its present state. The quest for that kind of all-encompassing explanation, known variously as a Grand Unified Theory or Theory of Everything, will occupy physicists deep into the twenty-first century and probably well beyond.

That great endeavor will probe the most basic issues, including the nature of wave-particle duality (which has been observed and manipulated, but not fully understood), the fundamental symmetries that govern the physical world, and the ways in which those symmetries are "broken" at different energy levels and in different physical conditions.

In addition, physics now has the ability to create and study structures and states that were inaccessible only a few years ago. Scientists already can create atom-sized devices, called "quantum wells," that share individual electrons just as atoms do, and they can tease groups of atoms into forming surprising new molecular arrangements. The startling discovery late in the century of buckyballs, a heretofore unknown form of carbon (one of the best characterized elements in the history of science), suggests that there may be vastly more to learn about even the most familiar entities. New discoveries arrive almost daily about the characteristics of exotic materials such as thin films, liquid crystals, and tiny solid-state lasers.

Further developments will enable researchers to investigate the behavior of currents, fields, lattice patterns, and vibrations at astonishingly small scale—and to uncover the profound geometrical beauty that so frequently accompanies, and so often helps explain, nature's assemblies.

Similarly, newly uncovered states of matter offer new research frontiers. Laboratory apparatus can now chill matter to the point at which its constituent atoms behave as a single collective quantum entity, opening an entirely new world to scientific scrutiny that will not be exhausted for decades. Physics has yet to achieve the lowest possible temperature, or the kinds of titanic pressures that are predicted to turn hydrogen gas into a stable metallic state—as many scientists believe occurs at the center of Jupiter and other giant outer planets.

Nor are physicists any longer obliged to rely only on the somewhat limited set of chemical combinations and molecular structures provided naturally by fewer than one hundred elements. As understanding of bonds and solid-matter configurations improves, researchers will be increasingly able to devise "designer" molecules with specific optical or other properties.

At the same time, entirely new patterns of natural order may be revealed as physicists improve their comprehension of "nonlinear" systems, thanks to the nascent sciences of chaos and complexity. Many phenomena that now seem largely inscrutable—from the rhythms of the human heart to the turbulence of fluids and the vagaries of weather—may yield to methods of analysis still in their infancy.

As progress continues, many more physicists and policymakers will be struck by the growing discontinuity between the sophistication of scientific understanding and the comparative crudity of our most common technologies. Indeed, one of the great challenges of the coming century will be to capitalize on the processes that modern science has uncovered. There is, after all, no shortage of practical matters begging for clarification—and soon.

No one seriously doubts that civilization will eventually have to find a substitute for fossil fuels. They powered the industrial revolution and still provide the overwhelming energy resources for modern society. But their future is inherently limited because of both diminishing resources and increasing concern about their effect on global climate. Physics must begin to provide longer-range solutions, whether by improving photovoltaics (which use sunlight, the ultimate source of nearly all the energy on Earth, to generate electricity), harnessing the vast power released in nuclear fission and fusion, exploiting the energy in noncombustion chemical reactions, or inventing other methods unimagined at present.

At the same time, physicists need to understand the nature of superconductivity in far better detail. Finding or creating materials that are superconducting at relatively high temperatures would have an enormous effect on daily life through resistance-free power transmission and vastly more efficient motors and generators. Many experts believe that a dependable room-temperature superconductor would have the potential to transform modern society as dramatically as did the initial introduction of electric power.

Many other practical advances are awaiting progress in basic research. One is the ongoing quest to obtain real-time images of the way living things function. Nuclear magnetic resonance and positron emission are only the beginning of this effort, which may culminate not only in drastically enhanced understanding of organisms and the physico-chemical processes of life itself, but in the creation of synthetic substances that can mimic—and perhaps even improve on—the systems that arose through evolution.

That work will require not only images with much finer resolution, but the ability to control the duration of light sources with even greater precision. Just as it takes a high-speed camera shutter to capture the frantic action of a soccer game, it will require laser pulses around a quadrillionth of a second to begin to get accurate snapshots of chemical reactions as they happen.

Another area of urgent concern also involves control of light. The conventional computer circuits on which civilization increasingly depends are about to reach a limit of miniaturization. Existing technology offers no way to create workable circuits thinner than 0.1 micron (about 1/500 the width of a human hair), and scientists are beginning to look for alternatives to carrying and processing information electronically. Optical computers are an attractive option; but building them will entail a much greater understanding of how to manipulate the optical properties of various kinds of materials.

Meanwhile, research will continue into the deepest questions concerning our place in the solar system and in the cosmos at large. Geophysics will harness new technologies in an attempt to resolve questions that still remain stubbornly unanswered. What is the source of the Earth's magnetic field and why does it change over geologic time? What, exactly, is the chemical makeup of the deep structure of the planet? What constraints govern the way planets form and the probability of different types? And what can that understanding tell us about the likely composition of surface and atmosphere on other worlds that are our inevitable frontier?

At the same time, the fruitful symbiosis between particle physics and astrophysics will flourish, as scientists search for a description of the invisible but essential "dark matter" that holds the universe together and accounts for at least 90 percent of its mass. Ongoing investigation of antimatter, including the eventual construction and study of entire "anti-atoms," will shed light on one of the strangest aspects of nature as we know it. That is, why there is a baffling preponderance of ordinary matter over antimatter in a cosmos that should have large quantities of both.

Nature has by no means exhausted its marvels. And all indications suggest that the next hundred years, perhaps even more than our present age, will be a century of wonder.

ACKNOWLEDGMENTS

We want specially to thank the members of the Editorial Advisory Committee, listed below, each of whom read portions of the text and made very helpful comments. Many individuals helped locate worthy images and, while we could not use all of them, we thank them all for their assistance. Alison Marcotte and Erika Ridgway played central roles in the preparation of this book. We thank Alison for her efforts to find and document the many images and Erika for keeping us all on schedule and on task, as well as Malcolm Tarlton, who created the diagrams. Finally, we want to express our appreciation for the efforts of Eric Himmel, our editor at Harry N. Abrams, Inc., who provided critical assistance in planning the book and at every stage of its development and worked with the three of us with great tact and good humor, and to Ellen Nygaard Ford, the designer, for her help in showing the beauty and excitement of physics.

Judy Franz, John Rigden, and Curt Suplee

Editorial Advisory Committee:
Jerome Friedman, MIT
Ernest Henley, University of Washington, Emeritus
Philip Morrison, MIT, Emeritus
George Pake, Institute for Research on Learning; Xerox, retired
Norman Ramsey, Harvard University, Emeritus
Arthur Schawlow, Stanford University, Emeritus
Roland W. Schmitt, General Electric, Retired; Rensselaer Polytechnic
    Institute, Emeritus
Frederick Seitz, Rockefeller University, Emeritus
Andrew Sessler, Lawrence Berkeley National Laboratory
Roger Stuewer, University of Minnesota
Spencer Weart, American Institute of Physics

INDEX

Page numbers in *italics* refer to illustrations.

PHOTOGRAPH CREDITS

Page 7: Courtesy of AIP Emilio Segre Visual Archives, Niels Bohr Library

Page 11: Courtesy of John Wagner, Ann Arbor, Michigan

Page 13: Courtesy of California Institute of Technology Archives. All rights reserved

Page 14 (both): Courtesy of The Cavendish Laboratory, University of Cambridge, England. Thanks to Miss H. M. Coote, Department Head, and Shirley Fieldhouse, Administrative Aide. Copyright

Page 15: Courtesy of Stanford Linear Accelerator Center (SLAC), Stanford University. Thanks to Dr. P. A. Moore, Assistant to the Director of Public Affairs

Page 17: Photo by Paul Harteck, courtesy of AIP Emilio Segre Visual Archives, Niels Bohr Library

Page 20: Courtesy of AIP Emilio Segre Visual Archives, Niels Bohr Library

Page 21: Courtesy of Brookhaven National Laboratory, Upton, New York. Thanks to Dawn Mosoff, Sr. Public Affairs Assistant

Page 23 (top): Courtesy of J. W. Harrell, Professor and Chairperson, Department of Physics and Astronomy, University of Alabama

Page 23 (bottom) and 25 (top): Courtesy of Lawrence Berkeley National Laboratory, Berkeley. Thanks to Marilyn Wong, Photography and Digital Imaging Services

Page 25 (bottom): Courtesy of Fermi National Accelerator Laboratory, Batavia, Illinois. Thanks to Judy Jackson, Public Affairs

Page 26: Courtesy of Department of Physics, University of Toronto

Page 27: Courtesy of Omicron Vacuum Physics, Inc.

Page 28: Electron micrograph by Lennart Nilsson, Stockholm. Courtesy of Surgical Medicine, New York

Page 29: "Bluegenes #1" © 1983 with all rights reserved by DesignerGenes Posters Ltd., P. O. Box 100, Del Mar, CA 92014 in Trust for the Julius Marmur Memorial Fund, Albert Einstein College of Medicine, Bronx, NY 10461

Page 30 (top left): Photograph by Erwin Mueller, Professor of Physics, courtesy of Pennsylvania State University

Page 30 (top right): Courtesy of Prof. T. T. Tsong, Distinguished Professor of Physics Emeritus, The Pennsylvania State University, and Distinguished Research Fellow and Director, Academia Senica, Taiwan

Page 30 (bottom): From The Double Helix by James D. Watson, Atheneum Press, N.Y., 1968

Page 31: Courtesy of Jeramia Ory, Department of Biochemistry, University of Minnesota

Page 32 (both): Courtesy of David Wineland, National Institute of Standards and Technology

Page 33: Courtesy of Mark Helfer, National Institute of Standards and Technology, Gaithersburg, Maryland

Page 34 (top): Courtesy of Donald M. Eigler, IBM Fellow, IBM Almaden Research Division, San Jose, California

Page 34 (bottom): Courtesy of Patricia Molinas-Mata, France's Commission for Atomic Energy

Page 35: Courtesy of Naval Research Laboratory, Washington, D.C. Thanks to Linda Greenway, Exhibits Manager

Page 36: Photo by Research Corporation, courtesy of AIP Emilio Segre Visual Archives, Niels Bohr Library

Page 37: Courtesy of Robert V. Pound, Department of Physics, Harvard University

Pages 38-39: Courtesy of Dale Mann, General Electric Research and Development Center, Schenectady, New York

Page 41: Courtesy of AIP Emilio Segre Visual Archives, Niels Bohr Library

Page 42 (top left and right): Courtesy of P. D. Weidman, Department of Mechanical Engineering, University of Colorado, Boulder, and V. O. Afenchenko, A. B. Ezersky, S. V. Kiyashko, and M. I. Rabinovich, Institute of Applied Physics, Russian Academy of Science, 46 Uljanov Str., 603600 Nizhny Novgorod, Russia

Page 44: University of Kentucky, courtesy of AIP Emilio Segre Visual Archives, Niels Bohr Library

Page 46 (both): Courtesy of Catherine Lawson, Brookhaven National Laboratory, Upton, New York

Page 47: Courtesy of C. D. Lima and W. A. Hendrickson, College of Physicians and Surgeons of Columbia University

Page 48: Photo by Yohkoh mission of ISAS, Japan, courtesy of Jennifer Conan-Tice, Stanford University

Page 49: Courtesy of J. T. Trauger, Jet Propulsion Laboratory, Pasadena, and NASA

Page 50 (both): Courtesy of Linda Hermans, Infrared Processing and Analysis Center, Caltech/Jet Propulsion Laboratory

Page 51: Courtesy of Naval Research Laboratory, Washington, D.C. Thanks to Linda Greenway, Exhibits Manager

Pages 54-56: Courtesy of Massachusetts Institute of Technology Museum. All rights reserved

Page 57: Courtesy of National Institute of Standards and Technology, Boulder, Colorado

Page 60 (top): Courtesy of David Weeks, Hughes Research Laboratories, LLC, Malibu, California

Page 60 (bottom): Courtesy of Charles Townes, Department of Physics, University of California, Berkeley

Page 61 (top): Courtesy of Spectra Physics, Mountain View, California

Page 61 (bottom): Photo by Steven Chu, Tom Perkins, and Doug Smith, Stanford University. Reprinted with permission from Science 264 (6 May 1994). Copyright 1994 American Association for the Advancement of Science

Page 61 (right): Courtesy of IBM Research. Thanks to James Wynne, Program Manager, Local Education Outreach, IBM

Page 63: Courtesy of Lucent Technologies Bell Labs Innovations, Murray Hill, New Jersey. Thanks to Jennifer Hammond, Public Relations

Page 65: Courtesy of Caltech Archives. All rights reserved

Page 66 (both): Courtesy of AIP Emilio Segre Visual Archives, Niels Bohr Library

Page 67: Courtesy of Wolfgang Ketterle, Massachusetts Institute of Technology

Page 68 (top): Courtesy of Michael Matthews, JILA, University of Colorado at Boulder

Pages 68 (bottom) and 69: Courtesy of Wolfgang Ketterle, Massachusetts Institute of Technology

Page 70 (both): Courtesy of AIP Emilio Segre Visual Archives, Niels Bohr Institute

Page 71 (top): Courtesy of AIP Emilio Segre Visual Archives Lande Collection, Niels Bohr Library

Page 71 (center): Photo by Paul Ehrenfest Jr., courtesy AIP Emilio Segre Visual Archives, Niels Bohr Library

Page 71 (bottom): Courtesy of AIP Emilio Segre Visual Archives, gift of Mrs. Mark Zemansky

Pages 74-75: Photos submitted by E. Ward Plummer, University of Tennessee, Knoxville, and Oak Ridge National Laboratory; B. G. Briner, P. Hofmann, M.Doering, H. P. Rust, A. M. Bradshaw of the Fritz-Haber-Institut der Max-Planck-Gesellschaft, Berlin; and L. Petersen, P. Sprunger, E. Laegsgaard, F. Besenbacher of the Institute of Physics and Astronomy, University of Aarhus, Denmark

Page 78: Photo by Carl D. Anderson, courtesy of Caltech Archives. All rights reserved

Page 79: Courtesy of Caltech Archives. All rights reserved

Page 82: Courtesy of AIP Emilio Segre Visual Archives, Niels Bohr Library

Page 84: Courtesy of Prof. R. E. Packard, Department of Physics, University of California, Berkeley

Page 85: Courtesy of Mark Gulley, Los Alamos National Laboratory

Pages 86-87: Courtesy of Donald M. Eigler, IBM Fellow, IBM Almaden Research Division, San Jose, California

Page 89: Courtesy of Naval Research Laboratory, Washington, D.C. Thanks to Linda Greenway, Exhibits Manager

Page 90: Courtesy of Hall of Electrical History of the Schenectady Museum, Schenectady, New York. Thanks to John Anderson and Joan Ahearn

Page 91: Courtesy of Massachusetts Institute of Technology Museum. All rights reserved

Page 95 (top): Courtesy of Lucent Technologies Bell Labs, Murray Hill, New Jersey

Page 95 (bottom): Courtesy of Jack S. Kilby, Texas Instruments

Page 96: Courtesy of Texas Instruments

Page 97: Courtesy of Iowa State University. Thanks to John McCarroll, University Relations

Pages 98 and 99 (top): Courtesy of Intel Museum Archives and Collection, Santa Clara, California

Page 99 (bottom): Courtesy of Marc Kastner, Massachusetts Institute of Technology

Page 100: Courtesy of TRW. Thanks to Jay McCaffrey, Director Public Relations

Page 101 (both) Courtesy of General Motors Global Research and Development Operations

Page 102: Courtesy of J. Unguris, R. J. Celotta, M. R. Scheinfein, D. T. Pierce, National Institute of Standards and Technology

Page 103: Courtesy of National Institute of Standards and Technology and Allied Signal Corporation

Page 104: Courtesy of Robert Celotta and Daniel Pierce, National Institute of Standards and Technology

Page 105: Courtesy of E. D. Dahlberg, Magnetic Microscopy Center, University of Minnesota

Page 106 (top): Courtesy of Max G. Lagally, University of Wisconsin, Madison

Page 106 (bottom): Courtesy of Omicron Vacuum Physics, Inc.

Page 107 (top): Courtesy of Paul Steinhardt, Princeton University

Page 107 (bottom): Courtesy of National Institute of Standards and Technology, Electron Physics Group

Page 108: Courtesy of J. F. Allen, University of St. Andrews, Scotland

Page 109: Courtesy of ISIS Facility, Rutherford Appleton Laboratory, United KIngdom

Page 112: Courtesy of General Electric Corporate Research and Development

Page 113: Courtesy of Royal Philips Electronics N. V., The Netherlands

Page 114: Courtesy of Naval Research Laboratory, Washington, D.C. Thanks to Linda Greenway, Exhibits Manager

Page 115: Courtesy of Michael W. Davidson, National High Magnetic Field Lab, Florida State University

Pages 116-117: Photos by Dr. D. J. Byron, Nottingham Polytechnic, courtesy of Institute of Physics, London.

Page 118 (both): Courtesy of Sandia National Laboratories Intelligent Micromachine Initiative

Page 119: Courtesy of A. N. Cleland, Department of Physics, University of California Santa Barbara, and M. L.

Roukes, California Institute of Technology

Page 121: Courtesy CERN Laboratories Media Service

Page 124: Courtesy of Brookhaven National Laboratory. Thanks to Dawn Mosoff, Sr. Public Affairs Assistant

Page 125: Courtesy of ISIS Facility, Rutherford Appleton Laboratory, United Kingdom

Page 127 (left): Courtesy of Al Wattenburg and Argonne National Laboratory

Page 127 (right): Courtesy of General Atomics, San Diego. Thanks to Doug Fouquet, Public Relations

Page 128: Courtesy of Caltech Archives. All rights reserved

Page 129 (all): Courtesy of Los Alamos National Laboratory

Page 131: Courtesy of General Atomics, San Diego. Thanks to Doug Fouquet, Public Relations

Pages 132-133: Courtesy of Princeton Plasma Physics Laboratory. Thanks to Anthony R. DeMeo, Information Services

Pages 134-135: Courtesy of Jeff Quintenz, Deputy Director, Pulsed Power Sciences, Sandia National Laboratory

Page 136: Courtesy of Stanford Linear Accelerator Center (SLAC). Thanks to Dr. P. A. Moore, Public Affairs

Page 137: Courtesy of Brookhaven National Laboratory. Thanks to Dawn Mosoff, Sr. Public Affairs Assistant

Page 138 (top): Courtesy of Peter Yamin, Brookhaven National Laboratory

Page 138 (bottom): Courtesy of Brookhaven National Laboratory. Thanks to Dawn Mosoff, Sr. Public Affairs Assistant

Pages 139-140: Courtesy of David Ellis, Thomas Jefferson National Accelerator Facility, Newport News, Virginia

Page 141: Courtesy of Fermi National Accelerator Laboratory, Batavia, Illinois. Thanks to Judy Jackson, Public Affairs

Page 142: Courtesy of Lawrence Berkeley National Laboratory, Berkeley. Thanks to Marilyn Wong, Photography and Digital Imaging Services

Page 143: Courtesy of Fermi National Accelerator Laboratory, Batavia, Illinois. Thanks to Judy Jackson, Public Affairs

Page 144: Courtesy of Renilde Vanden Broeck, CERN

Page 146 (both): Courtesy of Lawrence Berkeley National Laboratory, Berkeley. Thanks to Marilyn Wong, Photography and Digital Imaging Services

Page 147 (top): Courtesy of Fermi National Accelerator Laboratory, Batavia, Illinois. Thanks to Judy Jackson, Public Affairs

Page 148 (bottom): Courtesy of Brookhaven National Laboratory. Thanks to Dawn Mosoff, Sr. Public Affairs Assistant

Page 149: NA49 Collaboration, CERN

Pages 150-151: Courtesy of Fermi National Accelerator Laboratory, Batavia, Illinois. Thanks to Judy Jackson, Public Affairs

Page 153: Courtesy of Maarten Rutgers and Xiao Lun Wu, Ohio State University

Page 154: Courtesy of Akira Yoshida, Tokyo Denki University

Page 155 (both) Courtesy of Prof. Hiroshi Higuchi, Syracuse University

Pages 157-159: Courtesy of K. Y. Hsu, L. D. Chen (U. Of Iowa); V. R. Katta, L. P. Goss, D. D. Trump (Systems Research Labs); W. M. Roquemore (Wright Laboratory, USAF)

Pages 159-160: Courtesy of W. M. Roquemore, R. L.Britton. Air Force Research Laboratory

Page 162: Courtesy of James Wynne, Local Education Outreach, IBM Research

Page 163: Courtesy of General Atomics, San Diego. Thanks to Doug Fouquet, Public Relations

Page 165: Courtesy of Naval Research Laboratory, Washington, D.C. Thanks to Linda Greenway, Exhibits Manager

Page 166: Courtesy of Maarten A. Rutgers, Department of Physics, Ohio State University

Page 167: Courtesy of James Duncan, University of Maryland

Page 169: Courtesy of Jim Stone, University of Maryland

Page 171: Courtesy of P. B. Umbanhowar and H. L. Swinney, University of Texas at Austin, Center for Nonlinear Dynamics. Copyright 1996

Page 172: Courtesy of C. Fred Driscoll, University of California, San Diego

Page 173 (top): Courtesy of E. F. Hasselbrink, L. Muniz, M. G. Mungal, Stanford University Mechanical Engineering Department

Page 173 (bottom): Courtesy of A. Sonnenschein, P. Tuffy, University of Edinburgh. Courtesy IOP

Page 174: Courtesy of T. W. Ebbesen, H. J. Lezec, H.Hiura, J. W. Bennett, H.

F. Ghaemi, T. Thio, *Nature* 382, 52-56 (1996)

Page 175: Courtesy of R. E. Smalley, Rice University

Page 177: Courtesy of J. F. Allen, St. Andrews University, Scotland

Page 178: Courtesy of M. Frank G. Johnson, Northwestern University

Page 181: Courtesy of the University of New Hampshire. All rights reserved

Page 184 (both): Courtesy of Robert V. Pound, Harvard University

Page 185: Courtesy of California Institute of Technology

Pages 186-187: Courtesy of Charlotte Harris, McDonald Observatory, University of Texas at Austin

Page 188: Courtesy of Wendy Freedman, Carnegie Observatories

Page 189: Courtesy of R. Elson and R. Sword (Cambridge University) and NASA

Page 191: Courtesy of Naval Research Laboratory, Washington, D.C. Thanks to Linda Greenway, Exhibits Manager

Page 192: AIP Emilio Segre Visual Archives, Physics Today collection, Niels Bohr Library

Page 195: After de Lapparent, Geller and Huchra 1986 (Astrophysical Journal Letters); graphics by M. Kurtz. Copyright 1997 Smithsonian Astrophysical Observatory

Page 196: Courtesy of NASA's Goddard Space Flight Center and the COBE Science Working Group

Page 199: Courtesy of Lockheed Missiles and Space Co., Inc.

Page 200: Courtesy of University of Tokyo Institute for Cosmic Ray Research

Page 202: Courtesy of Space Telescope Science Institute

Page 203: Courtesy of J. Hester and P. Scowen (Arizona State University), and NASA

Page 204: Courtesy of M. Bobrowsky of Orbital Sciences Corp., and NASA

Page 205: Courtesy of H. Bond (STSCI), B. Balick (University Of Washington), and NASA

Pages 206-208: Courtesy of David Malin, Anglo-Australian Observatory

Pages 210-211: Courtesy of Dr. C. Robert O'Dell, Space Telescope Science Institute

Page 212: Courtesy R. Williams (STSCI), the Hubble Deep Field team, and NASA

Page 213: Courtesy J. M. Colberg, MPI fuer Astrophysik, Garching, and the VIRGO Consortium

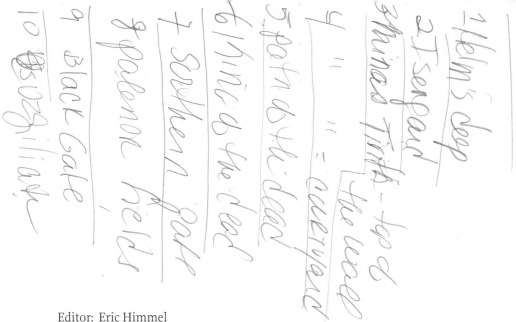

Editor: Eric Himmel
Designer: Ellen Nygaard Ford

The Library of Congress has cataloged the clothbound edition as follows:

Suplee, Curt.
    Physics in the twentieth century / Curt Suplee ; edited by Judy R. Franz and John S. Rigden
              p.   cm.
    Includes index.
    ISBN 0–8109–4364–6
    1. Physics—Popular works. I. Franz, Judy R. II. Rigden, John S.   III. Title.
  QC24.5.S86 1999
  530—dc21   98–41306
    ISBN 0-8109-9084-9 (paperback)

Printed and bound in Japan
10 9 8 7 6 5 4 3 2 1

Harry N. Abrams, Inc.
100 Fifth Avenue
New York, N.Y. 10011
www.abramsbooks.com

Abrams is a subsidiary of

LA MARTINIÈRE
GROUPE